Responsible Drone Journalism

Camera drones provide unique visual perspectives and add new dimensions to storytelling and accountability in journalism. Simultaneously, the rapidly expanding uses of drones as advanced sensor platforms raise new legislative, ethical and transparency issues.

Responsible Drone Journalism investigates the opportunities and dilemmas of using drones for journalistic purposes in a global perspective. Drawing on a framework of responsible research and innovation (RRI), the book explores responsible drone journalism from multiple perspectives, including new cultures of learning, flying in lower airspace, drone education and concerns about autonomous agents and big data surveillance.

By widening the discussion of drone journalism, the book is ideal for journalism teachers and students, as well as politicians, lawmakers, drone developers and citizens with an interest in the responsible use of camera drones.

Astrid Gynnild is Professor and Principal Investigator of the ViSmedia project at the University of Bergen, Norway. She is also Head of the Journalism Program, Department of Information Science and Media Studies, at the University of Bergen. Gynnild has published widely in the field of journalism innovation, creativity and new visual technologies.

Turo Uskali is Senior Research Scholar and heads the Journalism Program, Department of Language and Communication Studies, at the University of Jyväskylä, Finland. Uskali has published internationally over 20 peer-reviewed articles and book chapters focusing on innovations in journalism.

Disruptions: Studies in Digital Journalism

Series editor: Bob Franklin

Disruptions refers to the radical changes provoked by the affordances of digital technologies that occur at a pace and on a scale that disrupts settled understandings and traditional ways of creating value, interacting and communicating both socially and professionally. The consequences for digital journalism involve far reaching changes to business models, professional practices, roles, ethics, products and even challenges to the accepted definitions and understandings of journalism. For Digital Journalism Studies, the field of academic inquiry which explores and examines digital journalism, disruption results in paradigmatic and tectonic shifts in scholarly concerns. It prompts reconsideration of research methods, theoretical analyses and responses (oppositional and consensual) to such changes, which have been described as being akin to 'a moment of mind blowing uncertainty'.

Routledge's new book series, ***Disruptions: Studies in Digital Journalism***, seeks to capture, examine and analyze these moments of exciting and explosive professional and scholarly innovation which characterize developments in the day-to-day practice of journalism in an age of digital media, and which are articulated in the newly emerging academic discipline of Digital Journalism Studies.

Hyperlocal Journalism and Digital Disruptions
Scott Downman and Richard Murray

Fake News
Falsehood, Fabrication and Fantasy in Journalism
Brian McNair

Journalism Design
Interactive Technologies and the Future of Storytelling
Skye Doherty

Responsible Drone Journalism
Edited by Astrid Gynnild and Turo Uskali

https://www.routledge.com/Disruptions/book-series/DISRUPTDIGJOUR

Responsible Drone Journalism

Edited by Astrid Gynnild
and Turo Uskali

Routledge
Taylor & Francis Group
LONDON AND NEW YORK

First published 2018
by Routledge
2 Park Square, Milton Park, Abingdon, Oxon OX14 4RN

and by Routledge
711 Third Avenue, New York, NY 10017

Routledge is an imprint of the Taylor & Francis Group, an informa business

British Library Cataloguing-in-Publication Data
A catalogue record for this book is available from the British Library

Library of Congress Cataloging-in-Publication Data
Names: Gynnild, Astrid, 1959– editor. | Uskali, Turo editor.
Title: Responsible drone journalism / edited by Astrid Gynnild and Turo Uskali.
Description: London ; New York : Routledge, 2018. | Series: Disruptions: studies in digital journalism | Includes bibliographical references and index.
Identifiers: LCCN 2017055112 | ISBN 9781138059351 (hardback : alk. paper) | ISBN 9781315163659 (ebook)
Subjects: LCSH: Journalism—Technological innovations. | Drone aircraft.
Classification: LCC PN4784.T34 .R47 2018 | DDC 070.4/3—dc23
LC record available at https://lccn.loc.gov/2017055112

ISBN: 978-1-138-05935-1 (hbk)
ISBN: 978-1-315-16365-9 (ebk)

DOI: 10.4324/9781315163659

Typeset in Times New Roman

Contents

Contributors

David Goldberg is Senior Associate Research Fellow, Information Law and Policy Centre, Institute of Advanced Legal Studies, University of London; Senior Visiting Fellow, Institute of Computer and Communications Law in the Centre for Commercial Law Studies, Queen Mary, University of London and Adjunct Associate Professor of Law, Southwestern Law School, Los Angeles.

Frode Guribye is Associate Professor in Media Science at the Department of Information Science and Media Studies, University of Bergen, Norway. He is a co-investigator of the ViSmedia research project.

Astrid Gynnild is Professor and Principal Investigator of the ViSmedia project at the University of Bergen, Norway. She is also Head of the Journalism Program, Department of Information Science and Media Studies, at the University of Bergen. Gynnild has published widely in the field of journalism innovation, creativity and new visual technologies.

Deborah G. Johnson is the Anne Shirley Carter Olsson Professor Emeritus of Applied Ethics in the Science, Technology, and Society Program at the University of Virginia. Johnson is Professor II at the Department of Information Science and Media Studies, University of Bergen, Norway.

Lars Nyre is Professor of New Media, Journalism and Technology at the Department of Information Science and Media Studies, University of Bergen, Norway. Nyre is a co-investigator of the ViSmedia research project and leader of the Norwegian branch of INJECT.

Turo Uskali is Senior Research Scholar and heads the Journalism Program, Department of Language and Communication Studies, at the University of Jyväskylä, Finland. Uskali has published internationally over 20 peer-reviewed articles and book chapters focusing on innovations in journalism.

Acknowledgments

We would like to express our thanks to the many colleagues and students locally and around the world who have been involved in experiments, discussions and investigations on drone journalism during the work on this book.

Thanks to the enthusiastic team at Routledge for invaluable support during the process: Kitty Imbert, Jennifer Vennall, Margaret Ferrally, and Autumn Spalding (of Apex CoVantage), with special thanks to the editor above editors – Bob Franklin.

Special thanks also to our families for providing us time to write.

And indeed, special thanks to the Research Council of Norway, NFR, which funded most of the research for this book through the ViSmedia project.

1 What is responsible drone journalism?

Astrid Gynnild and Turo Uskali

Introduction

A drone is a flying vehicle that is remotely piloted or programmed to perform autonomous actions. In journalism, drones are often referred to as flying robots or camera drones. More formally, they are known as unmanned aerial vehicles (UAVs), unmanned aerial systems (UASs) or remotely piloted aircrafts (RPAs). News events such as protests, floods, fires, warfare and underwater operations exemplify only a snippet of what might possibly be covered in outstanding ways with unmanned aerial vehicles available to reporters.

Since 2011, news outlets on all continents have gradually embraced the options for disseminating imagery captured by camera drones. With drones being available to anyone who is interested, professional photographers and civilians have immersed themselves in a disruptive technology that is growing into a global, multibillion-dollar industry. With new opportunities for drone experimenting, hobby pilots and data techies from a multitude of backgrounds are also attracted to the drones. This attraction, in turn, encourages further exploration of drones as a newsgathering tool, although the experimenting sometimes appears to be prompted more by the possibilities of technology than the requirements of journalism.

In this book, we explore how the rapid expansion of dronalism – the process of doing drone journalism (Goldberg et al., 2013) – challenges established journalism at its roots. In particular, we investigate the opportunities and obstacles confronting what we have termed *responsible drone journalism*. The concept of responsible drone journalism merges responsible journalism with drone journalism. But, as we shall explore, it does more than that.

When collecting data for this book, we were immersed in the most stunning video captures, for instance, of San Francisco day and night: www.youtube.com/watch?v=0vJJ4-vgkUk. Turo experienced the joy of students who posted the first successful videos from a solo drone in the Finnish fields. We watched hours of video clips demonstrating that in the future, drone giants like Global Hawk might not be the worst autonomous warfare

DOI: 10.4324/9781315163659-1

vehicle, although it might spend more than 30 hours in the air. The growing investments in micro or even nano-drones, barely the size of insects, open up possibilities for new kinds of urban military attacks.

At this point in history, though, politicians and entrepreneurs across the globe seem to focus predominantly on the adventurous growth prospects of the emerging civilian drone industry. In a high-tech country like Norway, commercial and civil uses of drones are promoted as the country's new oil – products made for the international market. Drones are launched as the ultimate, complementary solution in fields as different as military espionage, electricity networks inspection and undersea iceberg identification. The Norwegian drone industry, for instance, currently employs 10,000 people but envisages employing nine times as many within three years.

Internationally, the first drone taxis are about to be available for public hire. Drone taxis might actually resolve some of the problems with traffic jams in densely populated cities. Drone taxis should be fairly easy to regulate in accordance with existing aviation rules. Due to their size, they are visible from a distance, and they create enough noise for people to become aware of their movements.

Multiple perspectivizing

In journalism, drone technology exemplifies what is defined as a disruptive innovation (Bower and Christensen, 1995; Christensen, 1997; Christensen and Raynor, 2003). With camera drones, reporters are not dependent on renting helicopters or cranes to get aerial imagery; even though camera drones in several respects are still inferior to mainstream technologies such as helicopters, drones are quicker and cheaper to use, and they can easily provide videos from areas that were previously visually inaccessible (Gynnild 2014b).

Existing research on drone journalism indicates that authors are typically concerned with ethical issues such as privacy and safety on behalf of journalists (Culver, 2014; Cruz Silva, 2016; Gynnild, 2014a; Tremayne & Clark, 2014). Empirical studies demonstrate how drone reporters are stretched between technological opportunities on the one hand and professional codes of conduct on the other. A growing community of drone startup enthusiasts, in contrast, extends the notions of what visual journalism is or should be (Giones & Brem, 2017); educators are increasingly grappling with unexpected issues when designing drone courses in higher education (Marron, 2013). The long-term consequences of this agency are particularly interesting especially since camera drones constantly put the limits of Press Freedom around the world to the test (Lauk et al., 2016).

In the larger picture of a drone society in the making, lawmakers and government officials grapple with intricately logistical problems of unmanned aerial vehicles in lower airspace; international aviation rules are contested by governmental differences in the ways military, commercial and civilian uses of drones are perceived, and thus how various regulatory mismatches ought to be aligned. Commercial uses of drones, of which journalism constitutes a small but crucial part of the puzzle, are instigated by surprisingly large national variations in the regulatory perceptions of privacy (Silva, 2016). In his examination of the relationship between technology and culture, Howley (2018) argues that "media discourse plays a decisive role in shaping these new technologies, understanding their applications in various spheres of human activity, and integrating them into everyday life" (p. xv).

Based on our investigation of civilian drones during the last seven years, we propose that such remotely controlled, unmanned aerial vehicles are soon going to be a natural and ubiquitous part of our lives. Our digitally structured, steered and surveilled society will have to learn how to relate to unmanned aerial vehicles of all sizes, whether we like it or not. It is a fact that these vehicles are used for a great variety of purposes. Drones are no longer a military tool used mostly by allied forces in the Middle and Far East. Whether we live in big cities in the West or in remote areas of the world, as humans and citizens we will most likely be prompted to engage with the flying robots in ways that would have been unthinkable only a short time ago.

Imagine heavy traffic and logistics to be resolved, not only in the streets but in the lower airspace, up to 120 meters above ground as well. Imagine drones becoming just as common as cars and motorcycles; imagine drones taking over human work in domains as diverse as power-line inspection, humanitarian relief and espionage.

Imagine civic drones passing borders with all kinds of packages – and potentially with people, too. Female activists fly contraceptives to sisters in need in Catholic countries, while criminals use drones to smuggle weapon and drugs. Drone sports events will be organized in local, national and international levels and drone taxis can fly you over the traffic jams. These examples are not future scenarios. They are reality, even though it might still take years to implement drones fully into societies' communication and transportation systems. Even though global society is still at an early stage of drone development, we have analyzed enough data to be convinced that drone technology alone might turn the global society, as we know it, upside down.

First, we take as a fact that drones are here to stay. Second, since we claim that the disruptive use of civic drones challenges established journalism at its roots, we wanted to discuss and propose the various ways that these challenges might be responsibly encountered and possibly overcome. With the pace demonstrated by the rapidly expanding global drone industry, there seems to be no way back. But there might be several ways forward.

The case of micro air vehicles (MAVs)

In the aftermath of the first wave of enthusiastic experimenting with camera drones in journalism, new questions arise, including this one: in what ways might, or should, news professionals relate constructively to future swarms of micro air vehicles (MAVs)? These mechanical insects and insect swarms are developed by the military and constructed to work in urban areas. This technology has already grown to the point where drone insects, individually or in swarms, can operate within and outside of buildings and be equipped with suction cups. They can crawl, climb and be airdropped or hand launched. Once in place, according to marketing videos from the US Air Force, these invisible machines can be on missions that last for days and weeks. They are supposed to be able to tap energy from power lines or harvest energy from sunlight and winds. According to the video clip, which is one of several carrying the same message, the system "remains robust even when GPS is unavailable." When operating in swarms, the MAVs can oversee or attack large areas in no time.

But anyone with some knowledge of MAVs knows that micro-drones might be capable of more than filming. They might carry chemicals or sensors that detect chemicals, depending on their mission as autonomous agents. Just like the larger drones, MAVs serve the goals established by humans. When MAVs operate individually and autonomously, though, these micro machines are invisible to the human eye and noiseless. They are too small to be regulated by aviation rules, but they could, for instance, easily fly into an office and glue themselves to a place under a president's desk. As advanced sensor platforms, even nano-drones and micro-drones are equipped for collecting multiple layers of data. They are easily integrated with other technologies and are thus capable of remotely piloted mass surveillance. It is likely that these tools are going to be adopted by journalists, too. Responsible drone journalism appears to be a complex field indeed.

As researchers and journalists engaged in dronalism, we are obliged to find out more about drone technologies and their potential uses. The time is over for the one approach to drone development that is either high-tech optimistic or high-tech pessimistic – either critical or constructive. The time is here to find out what is really going on within these polarities from multiple perspectives.

We firmly believe that journalists, along with researchers and educators, at this point in time have a crucial opening to explore, inform, influence and impact on the further direction and governance of the ongoing evolution of drones in a responsible manner. Even though the use of civic drones remains a predominantly local and hyperlocal phenomenon, drone issues are of increasing concern to humans nationwide as well as worldwide. But to steer the flying robots in wanted directions requires new kinds of insights in tandem with an informed willingness to act and to take new kinds of risks. Thus, we propose that the term responsible drone journalism has a double meaning.

The aim of this book is thus fourfold: 1) to provide a conceptual overview, along with "down-to-earth" illustrations/cases of the multifaceted uses of camera drones in journalism; 2) to discuss aviation laws and the regulatory challenges of dronalism; 3) to discuss ethical dilemmas and raise awareness about privacy, transparency and surveillance aspects of using drones as a journalistic tool; and 4) to report and discuss in what ways drone technologies might be responsibly incorporated into higher education.

The responsible research and innovation approach (RRI)

In this book, we therefore suggest that taking action in accordance with the ideas and tools of responsible research and innovation (RRI) is one way to go. RRI is a methodological framework that helps to facilitate the co-creational, collaborative resources of universities, industry, education and civic society (Owen et al., 2012; Stilgoe et al., 2013; von Schomberg, 2011). The approach is closely linked to Horizon 2020 and to governance research efforts to develop a responsible technological growth in countries within the European Union. The framework is rapidly spreading to other continents and countries as well.

Proponents of the RRI approach aim to find sustainable solutions to the grand challenges of our time by filling in what is referred to as the responsibility gap from the lack of governmental control in a free market. Thus, seen from a visionary journalism perspective, the theme in this book addresses the grand challenges of knowledge and of security in society through a drone lens.

It should be mentioned here that the research project that was the breeding ground for this book, ViSmedia, www.vismedia.org, is derived from a responsible research and innovation approach. In the ViSmedia project, it is our job as researchers to explore the ideas of the RRI framework and to investigate how they might be adopted and adapted in emerging fields such as drone journalism. It has taken a good deal of time to get on the inside of these ideas. And at the same time, we find that the responsibility aspect of journalism innovation does have much to offer.

In the seminal work on responsible research and innovation, Owen and Stilgoe, the most prominent spokespeople of responsible research and innovation, suggested that the RRI approach is built on four pillars (Owen et al., 2012; Stilgoe et al., 2013) for action. These pillars, as discussed by Stilgoe et al., are *anticipation, reflexivity, inclusion* and *responsiveness*. They point to different stages in responsible research and innovation processes, and require from the people involved that they, too, ask what-if questions at every stage of the process.

In a civic drone context, these four pillars might be considered idealistic requirements of responsible learning among stakeholders using and developing a technology. To work responsibly with technological innovations means that people involved should not only explore what is technologically possible to carry out; any innovational process should be accompanied by systematic reflections and deliberation on what might happen in a diversity of contexts. Anticipation in the form of foresight and scenario building plays an important role. The RRI approach prompts participants to reflect critically on the long-term consequences of their developmental actions and to identify unexpected issues that might surface on the way. The responsiveness dimension prompts participants to be flexible about changing course during any project, in response to the processes of ongoing reflection and deliberation.

The dilemma of governance not being able to control what individuals do with the new technologies actualizes the framework of responsible research and innovation. The so-called normative anchor points that should be reflected in the production processes of new technologies, according to RRI, are that the products should be ethically acceptable, that they should contribute to sustainable development and that they should be socially desirable (Owen et al., 2012) – whatever that means to journalism. These anchor points might seem appropriate and accurate at first glance, but they are challenging to define and live up to in practice. As defined by Stilgoe et al. (2013: 1570): "Responsible innovation means taking care of the future through collective stewardship of science and innovation in the present."

A requirement to researchers following the tenets of responsible research and innovation is, moreover, that the research is carried out in collaboration with a variety of stakeholders. Co-creation is meant to ensure that investigations are relevant and applicable to people involved and that the research efforts capture what is actually going on in a field. Advocating responsible research and innovation in the context of drone journalism means to engage actively in ongoing developmental processes of a visual technology in the making. That is, if not a grand challenge, at least a great challenge that needs the format of a book to be properly explored.

The double meaning of responsible drone journalism

So far, drone journalism has been associated predominantly with the use of camera drones as a newsgathering tool. More specifically, drone journalism is considered a simple means to provide imagery from above; it is a new feature of visual storytelling. This emerging use of camera drones by journalists was first described in blogs from the Poynter Institute of Journalism shortly after the turn of the millennium. In 2004, Larry Larsen suggested that journalists could start using a newly launched consumer level "Predator Spy Plane" to cover stories from the air, and a month later he mentioned a spy satellite that could be bought by reporters for a low price.

But even though the ideas for a new data collection tool were pointed out more than two decades ago, the term drone journalism appears to have emerged several years later.

In this book, we propose that, in the future, the term drone journalism will include the coverage of drones as a news beat just as much as it will refer to the visual data gathering. Huge resources are now being invested worldwide in innovative military and civilian use of drones. These investments signal that in the near future, drone robotics and autonomous agents might change society in currently unimaginable ways.

We propose that drone journalism is an emerging dimension of computational exploration in journalism (CEJ): the "innovative processing that occurs at the intersection between journalism and data technology" (Gynnild, 2014a, p. 715). Computational exploration in journalism refers to the experimental collection and selection as well as the dissemination of algorithmic data by new technologies. We further propose that in order for journalists to become invested in covering drones as a news beat, reporters should get a chance to experiment more with the technology. Journalists are more likely to open up for issues of innovation if they have carried out some kind of experiments themselves; journalism innovation leads to innovation journalism (Gynnild, 2014a).

The value of such direct experience, exemplified now by an explosive interest in drones as a journalistic tool, will hopefully manifest itself in a broader investigative coverage of drone technologies in general. The emergent uses of drone technologies might be seen as an early marker of a paradigm shift in which society moves from being human-centered to becoming steered by ideas of robotics, artificial intelligence and autonomous agents. As such, drone journalism is apt to highlight more than the concerns and processes of a single new technology in the making. At a larger scale, it might help to identify crucial issues to be considered and acted upon when available resources are increasingly being allotted to high-tech innovation in society.

An American organization, the Society of Professional Journalists, stresses that journalists are expected to seek the truth and report it, minimize harm, act independently and be accountable. These four aspects of journalism responsibility, and similar guidelines, are found in most Western countries. Up to this point in media history, responsible journalism has thus basically referred to journalistic codes of ethics. With fake news becoming the new buzzword after the US presidential election in 2016, the implications of responsible journalism indeed need to be more thoroughly investigated and explicated – not only to journalists but also to the general public.

So what is responsible drone journalism? Based on the research findings in this book, we offer a definition that takes into account the double meaning of the concept:

> First, *responsible drone journalism* refers to using drones as a journalistic tool in alignment with ethical and legal requirements, enhancing transparency and promoting informed reflection, deliberation and foresight among citizens. Second, *responsible drone journalism* refers to covering drones as a news beat by investigating implications of using drones in society as a whole.

It is time for journalists to do more than passively observe what is going on or adopt the perspectives of the industry. It is time to act on the observations. By asking critical, investigative "what if" questions on the outcomes and consequences of drone innovation, news media have a unique opportunity to influence a debate on drones that is still missing. What-if questions are open-ended with a built-in constructive and creative approach to problems. Such questions might help to uncover how the emergent challenges posed by drones might best be handled by society. Ideally, such constructive, thought-provoking approaches to phenomena are at the base of quality journalism. But way too often, in the constant flow of short-lived digital messaging, discussions about the outcomes of constructive and creative risk-taking actions with new technologies are lost, nonexistent or sometimes simply forgotten.

Drones as a ubiquitous tool

When new technologies are surfacing, mapping the field is the first step to stimulate foresight of what might happen at a later stage of development. To find out what is actually going on, data have to be empirically grounded to the extent that predications can be made. But mapping a field where the actors, products and legal regulations are in constant flux requires researchers to confront new challenges. It also highlights the need for moving from a descriptive to an analytical level in order to understand what is happening. To pay justice to these complexities, the reader will find that a number of qualitative approaches are used in this book, from descriptive case studies to innovation pedagogy, conceptual overviews and philosophical discussions.

So under such circumstances, what would responsible drone journalism in the making be like? What would it be capable of, for instance, if advanced drone technologies are used not to help, but to harm people – within the civic realm? These are the questions that will be investigated and discussed from multiple vantage points in this book.

New cultures of learning

Developing responsible drone journalism evidently exemplifies what Thomas and Brown (2011) have termed a new culture of learning. In this new culture, most actors involved, professionals as well as non-professionals, are within a realm of constant exploration of what was previously unthinkable, for instance, learning to build and fly drones through an Internet forum. We are talking about options for immersing not only oneself, but large communities, in new cultures of learning where the authors suggest that "the classroom as a model is replaced by learning environments in which digital media provide access to a rich source of information and play, and the processes that occur within those environments are integral to the results" (Thomas & Brown, 2011, pp. 37–38).

When dealing with new technologies, human openness and willingness to learn from taking risks is just as important as "handling" the tools at a technical level. In the new culture of learning, Thomas and Brown claim, participants do not learn so much *about* the world as they learn through engagement *within* the world. More specifically, in the new culture of learning "the point is to embrace what we don't know, come up with better questions about it, and continue asking those questions in order to learn more and more, both incrementally and exponentially" (p. 38).

This holistic approach to learning fits well with our own experiences. While academic research in dronalism is still quite limited, there is an impressive amount of empirical data available on the Internet. A multitude of learning experiments going on inside and outside of higher education are shared in Internet forums as well.

Journalism is often referred to as a signature institution. What emerges in journalism at an early stage is later adopted and adapted by other institutions in society. Typical examples are the early adoption of new technologies such as offset printing, mobile phones and digital cameras. Another more human example is the gradual, yet early, substitution of staff with freelancers and stringers. However, news organizations have tended to be followers or even laggards when it comes to innovative uses of simple Internet-related technologies. Social media such as YouTube, Facebook, Instagram, Snapchat and lately Jodel have been adopted by young people long before the news media got interested. At the same time, news media still seem to be in an early adopter position when it comes to camera technologies. They have typically been pioneers in using the latest equipment for news photography throughout the decades. Journalists tend to be pioneering actors in camera drones as well as in virtual reality exploration. And maybe that is perfectly reasonable given that such gadgets do require some financial investments. It took, for example, decades before mobile phones, such as the iPhone or Nokia Communicator were adopted by teenagers.

Whereas drone journalism is typically considered a disruptive means to provide images or video clips from above, so far there appears to be less interest among researchers as well as journalists in drone journalism as a news beat. Thus, most of the empirical data analyzed in this book is found not in the legacy media but elsewhere on the Internet. Based on these experiences, our advice is to search for open and closed forums in social media for more factual knowledge on drones.

The nexus of these four perspectives is the ubiquitous options for learning provided by the digital turn and the Internet. Thomas and Brown (2011) suggested that this new culture of learning is characterized primarily by learning in the collective. Following their argument, the acquisition of new knowledge in an emergent field such as drone technologies would be based on three principles. First, the authors claim that "the old ways of learning are *unable to keep up* with our rapidly changing world." Second, the new media platforms to a large extent facilitate peer-to-peer learning. Third, "peer-to-peer learning is amplified by emerging technologies that shape the *collective* nature of participation with those media" (Thomas & Brown, 2011, p. 50).

The authors point out that the fundamental difference between a collective and an ordinary community is that collectives cannot be passive in the same ways as communities can; Thomas and Brown (2011) claimed that whereas people in a community "learn in order to belong," people in a collective "belong in order to learn" (p. 52) At the same time, collectives do not have any centers and often very few rules; people are free to participate or not to engage in the collective whenever they wish.

In a previous study of journalism innovation that leads to innovation journalism, Gynnild (2014a) identified three different learning arenas for computational exploration in journalism: the newsroom approach, the research approach and the entrepreneurial approach. While these arenas were easy to distinguish when it came to further development of data journalism, the current study suggests that the collective learning that goes on in the drone field is qualitatively different. The fast pace in which formal and informal exploration of drones collapse into collective networks is striking. In a very short time, new collective networks resolve challenges that would previously have been very difficult to handle technologically. Simultaneously, of course, the rapid development of such technologies makes it very hard to control by legislative means locally as well as globally.

One might argue that responsible research and innovation is a typically European strategy for innovation. Silicon Valley's so-called free entrepreurial model for innovation is different in the sense that its starting point is a general reluctance to follow or obey any instructions from the government. Moreover, the Chinese authoritarian model for innovation represents yet another model.

Therein lies the challenge for dronalism as a newsgathering tool and as a news beat. In the words of Owen et al., "Responsible innovation requires a capacity to change shape or direction in response to stakeholder and public values and changing circumstances. . . . We must therefore consider how systems of innovation can be shaped so that they are as responsive as possible" (Stilgoe et al., 2013: 1570).

Implications of responsible drone journalism

In the following chapters, the implications of responsible drone journalism are up for debate. We first explore the phenomenon from global sensoring and lawmaking perspectives. Next, we discuss burning issues of societal transparency and surveillance followed by reporting from pioneering educative projects in drone journalism in different parts of the world. Finally, we mount the responsible drone journalism debate into a set of three tentatively interrelated scenarios.

In Chapter 2, Astrid Gynnild and Turo Uskali zoom in on drones as an aerial sensor platform in journalism. Airborne sensors provide journalists, drone operators and ordinary citizens with breathtaking opportunities for data collection and advanced storytelling. We ask in what ways is civic drone use reported in the media? What stories are told; what challenges are identified? In what ways do journalists experiment with the flying robots? The chapter provides an overview of recent trends and developments of drone journalism globally, and variations and threshold events across continents are discussed. The global history of drones as a disruptive journalistic tool is traced back to the Occupy Wall Street Movement in the US in 2011, when the activist Tim Pool and his friends managed to live-stream drone videos from inside the Occupy camps in New York.

In Chapter 3, David Goldberg dives into the recent changes and discussions on drones and aviation regulations in Europe and the US. The technological opportunities raise a host of regulatory, monitoring and logistic dilemmas that are waiting to be resolved. Goldberg focuses on what he calls two undernoted aspects of dronalism. Goldberg points out that newsgathering is protected under Article 10 of the European Convention of Human Rights. Even if it is remunerated, drone journalism is not an ordinary commercial activity. Goldberg also discusses the enforcement of the norms, whether by regulators, police and courts, as the norms are what really matter day-to-day for operators using their aircraft for dronalism.

In Chapter 4, Deborah G. Johnson and Astrid Gynnild bring the dilemmas of camera drones as autonomous agents to the fore. Controversial aspects of privacy, transparency and surveillance in journalism are discussed. Even though the drones as unmanned aerial vehicles are considered as autonomous agents, similar to autonomous cars and trains, there are humans operating behind the scenes. So what does the idea of unmanned vehicles actually imply? The chapter investigates the dilemmas of hidden or invisible human agents in journalistic storytelling and how their intent or purpose with drone actions best can be identified and understood.

In Chapter 5, Turo Uskali and Astrid Gynnild discuss practices and experiences of a pioneering Nordic journalism school in Finland that has systematically developed a drone journalism course for MA students. From there, the chapter extends to the evolution of drone journalism education at American universities. It emerges that an underlying vision of drone journalism education is to foster the building of innovative mindsets among journalism students. Further comparisons suggest that dronalism serves as an eye-opener to the core challenges of news journalism. The hands-on training requires from teachers to take on roles as peer-to-peer explorers, gardeners and player-coaches. Finally, the chapter suggests three models for drone journalism education.

In Chapter 6, Lars Nyre, Frode Guribye and Astrid Gynnild highlight the implications of introducing drones as a high-risk technology in higher education. A programmable camera drone was the main tool for a design experiment in a practical course at a Norwegian university. The pilot study suggests that the perceived risks of using the drone triggered students' creativity and willingness to explore the tool, whereas administrators and teachers were hesitant and careful to the extent that students' creative momentum was slowed down. The chapter discusses the relationship between technology, risk and learning, and proposes four learning principles that should characterize what the authors term responsible innovation pedagogy.

In Chapter 7, Turo Uskali and Astrid Gynnild propose three scenarios of responsible drone journalism in the light of the responsible research and innovation framework. The chapter sums up the implications of the previous chapters. The main variables are aviation regulations, learning environments and governance investments. The authors identify the use of satellites as a potential next step of drone journalism. For journalists to send satellites to the sky means that existing local and hyperlocal data gathering by drones might be replaced, or extended into, a much larger scope – robotic eyes from space.

The book is written for multiple audiences: journalists, journalism students, media researchers, technologists, politicians, lawmakers, drone developers, and citizens who grapple with the evolving and disruptive uses of civic drones. The edition springs from the ViSmedia project at the University of Bergen, with partners in Finland and the United States. One of the primary aims of the project is to contribute new insights regarding the grand challenges of knowledge and security in modern society.

References

Bower, J. L. & Christensen, C. M. (1995). Disruptive Technologies: Catching the Wave, *Harvard Business Review, January–February*.

Christensen, Clayton M. (1997). *The Innovator's Dilemma: When New Technologies Cause Great Firms to Fail*. Boston: Harvard Business School Press.

Christensen, C. M. & Michael E. Raynor. (2003). *The Innovator's Solution: Creating and Sustaining Successful Growth*. Boston: Harvard Business School Press.

Culver, K. B. (2014). From Battlefield to Newsroom: Ethical Implications of Drone Technology in Journalism. *Journal of Mass Media Ethics*, *29*(1), pp. 52–64. www.tandfonline.com/doi/abs/10.1080/08900523.2013.829679.

Giones, F., & Brem, A. (2017). From toys to tools: The co-evolution of technological and entrepreneurial developments in the drone industry. *Business Horizons*, *60*(6), pp. 875–884.

Goldberg, D., Corcoran, M., & Picard, R. G. (2013). Remotely Piloted Aircraft Systems and Journalism: Opportunities and Challenges of Drones in News Gathering. https://ora.ox.ac.uk/objects/uuid:a868f952-814d-4bf3-8cfa-9d58da904ee3.

Gynnild, A. (2014a). Journalism Innovation Leads to Innovation Journalism: The Impact of Computational Exploration on Changing Mindsets. *Journalism*, *15*(6), pp. 713–730. http://journals.sagepub.com/doi/abs/10.1177/1464884913486393.

Gynnild, A. (2014b). The Robot Eye Witness: Extending Visual Journalism through Drone Surveillance. *Digital Journalism*, *2*(3), pp. 334–343. www.tandfonline.com/doi/abs/10.1080/21670811.2014.883184.

Howley, K. (2018). *Drones: Media Discourse and the Public Imagination*. New York: Peter Lang.

Lauk, E., Uskali, T., Kuutti, H., & Hirvinen, H. (2016). Drone Journalism: The Newest Global Test of Press Freedom. In Carlsson, Ulla (Ed.). *Freedom of Expression and Media in Transition: Studies and Reflections in the Digital Age*, pp. 117–125. Gothenburg, Sweden: Nordicom.

Marron, M. B. (2013). Drones in Journalism Education, pp. 95–98. http://journals.sagepub.com/doi/full/10.1177/1077695813486973.

Owen, R., Macnaghten, P., & Stilgoe, J. (2012). Responsible Research and Innovation: From Science in Society to Science for Society, with Society. *Science and Public Policy*, *39*(6), pp. 751–760. https://academic.oup.com/spp/article-abstract/39/6/751/1620724.

Silva, J. C. (2016). Legal and Ethical State of Drone Journalism in Andean Community Countries. http://pucedspace.puce.edu.ec/handle/23000/1087.

Stilgoe, J., Owen, R., & Macnaghten, P. (2013). Developing a Framework for Responsible Innovation. *Research Policy, 42*(9), pp. 1568–1580. www.sciencedirect.com/science/article/pii/S0048733313000930.

Thomas, D., & Brown, J. S. (2011). *A New Culture of Learning: Cultivating the Imagination for a World of Constant Change*. Lexington, KY: CreateSpace.

Tremayne, M., & Clark, A. (2014). New Perspectives from the Sky: Unmanned Aerial Vehicles and Journalism. *Digital Journalism, 2*(2), pp. 232–246. www.tandfonline.com/doi/abs/10.1080/21670811.2013.805039.

Von Schomberg, R. (2011). Prospects for Technology Assessment in a framework of responsible research and innovation. In *Technikfolgen abschätzen lehren: Bildungspotenziale transdisziplinärer Methode*, pp. 39–61, Wiesbaden: Springer VS.

2 The first wave of drone journalism

From activist tool to global game changer

Astrid Gynnild and Turo Uskali

Introduction

The story of Paris Hilton filmed by a drone on the French Riviera in 2010 has become a classic. The event marks the beginning of the drone era in visual journalism; possibly for the first time in history, paparazzi succeeded in securing aerial images of a world-class celebrity without using a helicopter. The disruptive technology was primitive but lightweight, inexpensive and met the needs of the photographer there and then (Gynnild 2013; Tremayne & Clark 2013). Combined with the options for rapid diffusion provided by the Internet and wireless networks, the paparazzi had gained a new competitive advantage.

From an activist perspective, though, we claim that the real gate opener for drones as a journalistic tool goes back to 2011, when Tim Pool and his friends managed what the world had not yet seen; they started live-streaming drone videos from inside the Occupy Wall Street camp in New York. First, they built their own small drones by simple means while sharing every step of the process with an increasing number of online followers. The drones were easily controlled by a smartphone; by utilizing the newly launched U-stream technology, the young innovators were able to provide visual live reports, from the air, of what was actually going on inside the activist camp. The counter-power approach was closely followed by thousands of people on the Internet; it prompted much buzz and hundreds of headlines as the drones documented activities that were quite different from those presented by the establishment and the police that were watching the camp (Gynnild 2013).

The event exemplifies a fact repeatedly stressed by Manuel Castells (2009, 2012): as power relations are embedded in the institutions of society, creative actions of counter-power are likely to pop up outside of the institutions. The more aggressively the police reported from the protests at the camp, the more the activists engaged in finding new ways to document realities from their perspective.

DOI: 10.4324/9781315163659-2

Within weeks, the disruptive breakthrough of the young innovators in New York evolved into a worldwide activist initiative; suddenly, small, homemade drones with limited flying capacity reached out to audiences on all continents, and clips were sold from activists to established news organizations such as CNN. In November, riots in Warsaw, Poland, were recorded by anonymous activists and circulated via YouTube (Goldberg et al. 2013; Corcoran 2014). In December, demonstrations in Moscow, Russia, were documented by ten aerial still images shot by a drone and published on the citizen news site ridus.ru (Christoprudov 2011; Goldberg et al. 2013). Self-made camera drones were still quite expensive – up to several thousand dollars each (Martinelli 2011).

At this point, though, it was still difficult to imagine that only a few years later, the term drone would be on everyone's lips, that drones would evolve into advanced sensor platforms (Pitt 2014) used by governments as well as by leading news organizations, that a new global multibillion-dollar industry was in the making and that opportunities for new ways for the surveillance of society from above would suddenly continue to grow exponentially.

Empirical data and case approach

In this chapter, we identify and discuss seminal cases of journalism innovation and pioneering actions in this evolution. In what ways was journalism a global test-bed for drones as a disruptive innovation tool? What dilemmas were identified, and how were such dilemmas and obstacles encountered in the first wave of drone journalism? The overview is grounded in an analysis of dozens of cases on drone journalism development across the globe from 2010 to 2018. Based on these data, we suggest that within less than a decade, camera drones have evolved to become a game changer in global news journalism. A game changer is a "newly introduced element or factor that changes an existing situation or activity in a significant way" (Merriam-Webster's Collegiate Dictionary, 11th ed.). In this context, a game changer is a new tool and new professional practice that changes the ways that journalism is produced – often more efficiently and more transparently than earlier.

Since we started collecting data for this study, the number of online sites in which drone journalism is disseminated and discussed has grown exponentially. Hobbyists, activists and professional journalists have from the outset shared incredible amounts of expert knowledge via social media, blogs and in other online communities. The field is marked by a speedy viral diffusion of imagery and a wealth of new ideas for anyone to use. So far, though, surprisingly little *research* has been conducted on drones applied to journalism or on drones as a news beat.

In this study we are concerned primarily with the expansion of drones as a newsgathering tool. The issues that we address in this chapter therefore stem from multiple sets of empirical data: drone video clips, discussion forums, blogs, social media, conferences, hearings and legislative documents, in addition to extant research. Interestingly, the many posts on doing drone journalism far outweigh the posts on drones as a news beat. The learning processes of dronalism seems to have evoked a kind of collective connectivism (Siemens 2005) among enthusiasts that make online drone communities particularly interesting to study.

In this chapter we present a few main, recurring themes from a vast material mostly written by activists and journalists. As researchers investigating a multibillion-dollar technology and its practical implications for the global news industry, we are, of course, constantly faced with the challenge of data overload in an expansive virtual universe. Our aim, however, is not to provide a full descriptive overview of drone journalism incidents but to contextualize what surfaces when a disruptive technology changes established premises for visual news work. Based on the available data, we propose that within less than a decade camera drones have evolved from being a primitive tool for activists to becoming a game changer in visual journalism. At the same time, the fact that we have only accessed empirical data written in English and Nordic languages is a clear limitation of the study. The material is also biased in the sense that what makes the headlines is usually rare events, the unexpected and the surprising, or the otherwise exceptional moves that take place within a field such as drone journalism.

First controversies

In this study, it was tempting to turn again to seminal theories on innovation, diffusion and disruptive technologies, such as Rogers (1995) or Christensen (1997) for solid explanations of the drone evolution. As we worked and reworked the material, however, we were struck by the enthusiasm and playfulness of individuals that came out of the data. These game-like issues of exploring a new technology prompted reflections on lost opportunities of creative freedom in the business and the ability to carry out journalism in a state of mental surplus. Such vague but repeatedly surfacing data are partly explained by Castells (2012) who emphasized that social networking on the Internet provides new spaces of individual autonomy – beyond the control of governments and corporations that had previously monopolized channels of communication power. Actually, the many stories provided by reporters were so thrilling that we decided to incorporate a relatively large amount of concrete examples in the analysis.

When new technologies are adopted and adapted in a new field, the lack of relevant rules and regulations leads to much confusion among its actors. Law professor and media historian Tim Wu (2011) points out that such anarchistic periods characterized the introduction of the telegraph, radio and television, as well as the Internet and mobile technologies. In terms of innovations, a certain legislative time lag is usually to be expected. Therefore, lawyers and law schools are often in the forefront of handling such issues case by case. Of course, international aviation regulations and crime laws were valid and operative from the beginning. But as we shall see, even if drones were regulated by international aviation rules, early adopters were constantly exploring invisible and visible boundaries of the new tool.

In the very early years, from 2010 to 2012, news about the first camera drone controversies were circulated widely via online groups, blogs and news outlets. Some incidents even started yearlong controversies and legal battles; typically, battles prompted by hobbyists and journalism activists immersed in the new activity. In early 2012, for instance, an American drone hobbyist detected possible environmental problems by watching his own drone footage. The clip displayed dangerous waste in a river near a meat packing plant in Dallas, Texas, in the United States. After contacting the local authorities, investigations started. Consequently, the meat packing plant was closed for a year and a half (Mortimer 2012a; Wilonsky 2013).

After this widely circulated incident, many other activists started using drones for similar actions, which often created headlines. Also in 2012, for example, in the United States, an animal rights group's drone was shot down while the activists witnessed and recorded a live pigeon shooting. The drone was shot down in an act of revenge by the pigeon hunters. One activist commented that it was a very short flight. The shooters had hidden themselves in the woods and as soon as the machine was up to about 150' they started shooting" (Schroyer 2012). The local sheriff's department filed a malicious damage to property incident report by the animal rights group. The incident also received international press coverage (Thetandd.com 2012; Keneally 2012). In July 2013, the Federal Aviation Administration (FAA) warned the public against shooting guns at drones, stating, "Shooting at an unmanned aircraft could result in criminal or civil liability, just as would firing at a manned airplane." The FAA released the statement after a town in Colorado started granting "hunting permits" to shoot drones (Lowy 2013).

A typical trait of these cases is the pioneering role of activists who used drones to document and report on political issues such as demonstrations, environmental crimes and animal rights issues. The disruptive technology provided extended opportunities for connecting large communities of people with similar interests. Such citizen drone reporting has, nearly without exception, proved to be difficult to handle for governmental authorities. which have repeatedly called for stricter regulations. Identifying crowd size during riots and the scope of damage and devastation in the aftermath of natural catastrophes is not always wanted by the police and by governments.

The counter-strategy of governmental authorities is typically to issue temporary flight restrictions, TFL (Temporary Flight Limitations). This strategy is widespread at least in most Western countries. During the Ferguson unrest in the United States in August 2014, a TFL was issued by the FAA. The protests and riots began after a fatal shooting of an 18-year-old African-American youth by a white police officer, and continued for many months. The no-fly-zone restriction exemplified that whenever controversial events happen in the United States, the airspace might easily be blocked by authorities. In practice, this means that with drones available, the work of journalists is hindered by governmental authorities who impose temporary flight restrictions (Dronejournalismlab.org 2014; Perritt & Sprague 2017a, p. 195). In Norway and other Northern European countries, temporary flight restrictions are regularly issued during fires and accidents, with the argument being that safety work might be at risk if unmanned aerial vehicles fly into the area.

Technology evolution as infinite gaming

In his provocative book, *What Technology Wants* (2010), the technology philosopher Kevin Kelly claims that the main aim of technological evolution is to keep the game of possibilities open. After seven years of studying new technologies, Kelly suggests that the technium, as a whole, is a kind of living, natural system that has unconscious, long-term tendencies built into the system – tendencies that cannot be avoided or stopped. Subsequently, he suggests that for humans, adopting principles of proactivity and engagement are the only ways to steer or tame technology in wanted directions. Keeping the game of possibilities open, as Kelly suggested, implies that any technology will constantly move in directions that generate more options to humans: more opportunities, more connection, more diversity, more unity, more ubiquity and more thoughts. Additionally, and as a consequence, new technologies generate more problems, too.

According to Kelly, technologies constantly engage in changing roles in society. He sees them as physical manifestations of infinite gaming, in which individuals constantly seek the "minimum amount of technologies that will create the max amount of choices for oneself and others" (Kelly 2010, p. 352). Moreover, he points out that technology permanently engages science, innovation, education and pluralism that allow individual humans to generate and participate in a greater number of ideas. In that way, technology allows each person to do better, he claims. The engagement in constant auto-creation of new ideas means that when playing the infinite technology game, humans "explore every way to play the game and include all possible games and players to widen what is meant by playing" (Kelly 2010, p. 354).

There is a big difference between playing finite and infinite games. When applied to human behavior, engaging in finite games means that individuals or groups of people engage in activities in which the frames for the game are known beforehand: time, place, number of players and rule, and where the goal is to end up with a winner of the game. In infinite gaming, by contrast, there are no winners or any end to the game. So the goal of the technium game, then, is to keep playing to constantly expand and continuously discover more opportunities; in that respect, to humans, engaging in online activities, no matter what content or which direction, in reality means to engage in infinite technological gaming.

Kelly clearly belongs to the large group of tech-optimists who are more invested in understanding the opportunities of new technologies than in the obstacles. His theorizing is liberating in the sense that conceptually he cuts across layers of wires and wireless connectivity; at the core of it all he sees technology development as a systematic force in which we can choose to immerse ourselves – for good or bad. But we can't choose to close new technologies out. We have to learn how to live with them and constantly develop our ability to make responsible choices.

Early coverage of disasters

In the perspective of droning as part of an infinite technological game, the first wave of drone journalism implied increased opportunities for activists to document and disseminate their actions. Moreover, the technical evolution provided paparazzi with new ways to achieve valuable shots through simpler means. A third step was the increasing efficiency and new visual freedom that journalists gained when covering disasters with camera drones. Floods, forest fires, erupting volcanoes in Vanuatu and Iceland and a damaged nuclear power plant in Chernobyl, Ukraine, are just a few examples. Many drones have been lost in these hazardous environments, but as far we know, no journalists have been hurt or killed (Mackley 2012; Lam 2014; Schroyer 2014). Areas that were once considered too dangerous, too remote or inaccessible in other ways now lay open for journalistic conquering. And they opened up new cultures for individual learning, as proposed by Thomas and Brown (2011).

As early as June 2011, the competitive advantage of camera drones was proved through the amazing coverage of floods in Alabama and North Dakota in the United States. Typically enough, the drone filming was initiated not by established news media but by reporters in an entrepreneurial iPad publication called The Daily (Hill 2011). Among practitioners and scholars it seemed to be agreed that covering stories of disasters and other possible harmful environments from a distance was at the core of drone journalism tasks (Holton et al. 2015, p. 634; Gynnild 2013; Mullin 2016).

Philip Grossman, the senior director for media technology and strategy at The Weather Channel, who focused on recording Chernobyl 30 years after the world's worst nuclear disaster, said in an interview that

> By providing images from a different perspective, one is able to tell a more complete story. Each perspective (ground, tripod, slider, drone) provides a different way to tell a story. It's sort of like "triangulation" by providing various reference points one can figure out where they are.
> (Schroyer 2014)

In April 2015, after the devastating earthquakes of 7.8 magnitude in Nepal, a local drone user posted aerial videos of damaged buildings in the capital, Kathmandu, through social media platforms. After the videos went viral, international news media republished the footage when reporting on the aftermath of the earthquakes (Shammas 2015; Sky.com 2015). This incident alone was followed by a number of international news organizations that wanted to use their own drone journalism teams on the spot. The Associated Press was the first news agency to provide extensive self-shot material from Kathmandu (Imregi 2015).

Moreover, in 2016, the European migrant crisis, in the category of a major conflict's aftermath, offered considerable emotional drone footage starting from life jacket "graveyards" in Greek islands to long queues at Hungarian borders. Often, no voice-overs were needed to tell the story effectively (BBC.com 2016).

Dilemmas of drone war reporting

Mark Corcoran (2014, p. 1) identified military conflict as the first major category of what he called "hazardous news gathering assignments," whereas the two other categories were civil unrest and disaster coverage. In the following section we discuss recurring dilemmas of camera drones used for journalism coverage of military conflicts.

Eastern Ukraine and Syria as major military conflict areas became the first test beds for drone journalism in war zones. For the first time in history, two wars, in parallel, were documented with camera drones. The journalistic flipside of the coin was that the drone footage stemmed mainly from military organizations and was meant for propaganda purposes.

Nevertheless, these drone clips, mostly showing the vast destruction of the conflict areas, were widely published by international news media like CNN and the Guardian, and were also circulated via social media platforms like YouTube (Theguardian.com 2015; Walsh 2015; Vaux 2015; Postema 2015; Theguardian.com 2016). The footage served the role of robot eyewitnesses (Gynnild 2013) from above, albeit manipulated, which was a clear technological extension of human opportunities (Kelly 2010) for documenting the cruelties of war.

Journalistically, one can argue that war-related footage should always be examined critically because of its potential propagandistic aims (Postema 2015; Uskali & Lauk 2018). Prominent news organizations such as the BBC and the *New York Times* have systematically avoided the use of propagandistic drone footage from wars. Instead, they have deployed their own drone journalism teams to produce original footage from conflict zones, especially in Eastern Ukraine. According to Postema (2015), drone reporting is evidently an important competitive asset for leading news media when it comes to hunting for the best possible visuals:

> The BBC reporters got closer to the frontline than The New York Times photographer, and they had the visual evidence of the devastation. But sending a war reporter with a drone to the frontline means taking a huge risk. – The drone seems not to be used to substitute risky war reporting operations, but to match the competition, to obtain their own drone report.

The particular value of drones in war reporting is an argument frequently used by drone researchers (Gynnild 2013; Lauk et al. 2016; Uskali & Lauk 2018). Experts like BBC's Thomas Hannen have warned that "using them in conflict situations would be dangerous, both to journalists and troops," and added that there is "certainly no way that you could do it safely, because as soon as you fly one of these things above your head, you're immediately identifying where you are, both visibly and audibly, because they're very loud" (Collins 2014). Corcoran (2014, p. 23) also cautiously mentioned that radio links needed for controlling the drones would be "relatively easy to intercept and locate using basic military signals intelligence equipment." The world's smallest drone developed for military purposes, the Black Hornet, weighs 16 grams (half an ounce) and is equipped with night vision capabilities and infrared sensors that can live-stream video still images within a 1.6 km range. This micro unmanned aerial vehicle was first used by Western troops in Afghanistan but is now freely available for anyone to buy and illustrates some of the emergent options available to journalism as well.

Indeed using drones does not necessarily reduce all the risks of war reporting. The journalist is still often close to the front lines. Perhaps only the satellites could really improve the situation. For example, 3D models about the destroyed Donetsk airport in Eastern Ukraine could be constructed based on drone and satellite images (Schroyer 2015).

But as Corcoran (2014, p. 26) points out:

> Conflict reporting is not just about military embeds. Equally important is the civil story: aid distribution, medical treatment, refugees, investigation of human rights abuses. In this environment media drones should only be deployed with great sensitivity.

Toward ubiquitous use in large corporations

When categorizing the data material of this study, we started out with a relatively detailed timeline of drone journalism approaches. It quickly emerged from the data that similar to the evolution of other disruptive technologies, the drone tool was first tested out by individuals at a decent distance from established newsrooms – not so much by entrepreneurs as by devoted hobbyists and actionists. The patterns of diffusion identified by Rogers (1995) were evident: the spread of drone journalism went from the ground up, from small to large, from periphery to center, and from online news sites to mainstream television. The BBC News was among the first TV stations. They debuted with their first "flying camera" in October 2013, but actually more than three years after the first successful uses of drones by paparazzi.

The first short BBC clip was related to a news story about a new high-speed rail line. The BBC called the self-made "flying camera" a "hexacopter" (Westcot 2013). The same tendency to avoid the term drone was noticeable in experiments by other TV stations during the same period of time. Among them was Norwegian Broadcasting, as they, too, wanted to avoid using the term drone because of its military connotations. The short history of drone journalism, though, proved that the power of the term drone has outweighed all other attempts to establish unmanned aerial vehicles as a technology easily distinguishable from the term drones used in a military context.

In this first news story, the flying camera was used "to surprise the viewer." It first acted like "a person walking along with a camera on their shoulder," and then, suddenly, it flew away, showing aerial images of a train station. It took four hours to get the first 20-second drone piece "telly right." Also, the loud noise of the rotors caused some problems, and the team had to reduce the sound in post-production. The BBC marketed that "This machine is going to transform the way TV news looks in the future"

(BBC.news.com 2013; Westcott 2013). The next use of the BBC's fleet of three hexacopters was at Stonehenge (Collins 2014).

Before shooting the first clip with a drone, the BBC, as a huge film actor in society, demonstrated professional responsibility by establishing filming rules based on the British Civil Aviation Authority regulations. In practice this meant that they were not able to fly within 50 meters of a road or building, fly over crowds, fly 500 meters horizontally or 120 meters vertically from the pilot. They also agreed to have a flight plan before every takeoff. In addition, they built an extra safety layer, a GPS-based system on board that ensured that if the radio link broke down, the drone automatically would fly back from where it took off and land (Westcott 2013). Camera drones were mainly used for outdoor reporting, but of course they could also be used indoors. For example, BBC shot indoor footage of their new broadcasting house while it was still under construction (Schroyer 2013a). Interestingly, ABC News used drones for live broadcasting in 2014 in Canberra during the Australia Day ceremonies. Within one hour of broadcast, the drone went "live" 25 times.

The BBC News crew tested the limits of using their "flying cameras" abroad. During the World Economic Forum in Davos in January 2015, three BBC journalists were briefly detained and questioned by the Swiss police after they used a drone in Davos's no-fly zone (Halliday 2015). In similar vein, Qatar-based satellite news network Al-Jazeera's three journalists were arrested in Paris, France, in February 2015 because of illegally flying a camera drone near city landmarks during the night. Flying drones over Paris without a license is banned by law. Also operating drones during the night is illegal (BBC.com 2015b; WAN/Ifra 2015).

The first wave of drone journalism is evidently characterized by an explorative vulnerability to controversies and accidents. Journalists were mostly very careful user-testers even in a period when aerial or other regulations were nearly nonexistent in many countries. Journalists apparently knew that there was little space for mistakes. The World Association for Newspapers (WAN/Ifra), started to pay attention to drone journalism and emphasized free press rights and free speech rights when the first country-wide bans of drones for journalism emerged 2013–2014 (Pead 2014; Corcoran 2014, 28).

Sometimes a single drone incident could trigger a total ban, such as in Kenya: a drone was seen a few minutes before the president was to arrive at a stadium to celebrate Kenya's national day (Flanagan 2015; Jakarta Post 2015; Johnson 2015). In Thailand, the new drone regulations set by the military junta in 2015 prohibited citizens from using camera drones; authorities also had the final say over what was allowed – and what was not – for the news media (Greenwood 2015). Indeed, other authoritarian regimes such as Nepal, Indonesia and Kenya followed suit. The government in Nepal prohibited the use of drones within a week after the earthquakes in 2015. The drone pioneer, Professor Matt Waite commented on the trend:

> This Nepal earthquake was one of the first [news events in which] media were truly aware of drones and their power. And they came, and it immediately triggered a ban. And I'm worried that that's going to happen again.
>
> (Flanagan 2015)

Perhaps most surprisingly of all, the otherwise liberal Nordic country of Sweden prohibited all use of camera drones for more than a year. In October 2016, the Supreme Administrative Court of Sweden targeted all recreational and commercial users alike, with no exception for journalists (Teirstein 2016). Media companies and several trade associations in Sweden criticized the new rules; the Swedish government's representatives assured them that the rules would be reversed later in 2017 (Thelocal.se 2016). Finally, in August 2017 new, less prohibitive drone rules were introduced in Sweden. According to these new regulations, hobbyists no longer needed a license from the authorities. But anyone who uses camera drones professionally, for instance, journalists, needs to apply for a license, pay fees and insurance, and also report about their flights (Eriksson 2017).

In general, the first pieces of drone footage started to emerge regularly in mainstream arenas of journalism during 2012–2013. In Italy, for example, the case of the wrecked cruise ship Costa Concordia in January 2012 offered a good showpiece for local drone operators to enter foreign newscasts (Caputo 2013; Zavrsnik 2016, p. 223).

In Latin America, the early adopters of drones for journalism were the Brazilian *Folha de São Paulo* and Globo, which used drones to record protests against government spending and rising public transportation prices. The Peruvian *La Prensa de Peru* covered roadwork and road closures from the air; Salvadoran *La Prensa Gráfica* applied drones for election coverage; in Mexico, the Grupo Reforma documented the construction of Latin America's tallest skyscraper (Diep 2014; DronesSkycam 2013).

Cable News Network (CNN) used a drone for recording the devastation of Typhoon Hayian in the Philippines in November 2013. It was one of the most intense tropical storms on record, and the aerial coverage got much attention in social media. In the story, the reporter, Karl Penhaul, made a stand-up in the middle of rubbish while the drone flew over him into the sky to show the magnitude of the devastation (Penhaul 2013).

Major news organizations in the United States started to explore the capabilities of drones in 2014–2015. First, NBC News used drones abroad for foreign news reporting from Vanuatu after a cyclone hit the island

(NBCnews.com 2015). The *New York Times* used drone footage for an environmental story about melting Greenland (Haner 2015). In addition, in 2015, the Federal Aviation Administration (FAA) granted some exemptions for the news media to allow them to fly drones inside the United States. For example, the Associated Press (AP) trained its first licensed drone pilots and started "experimenting with drones across videos and photos, but not yet on a regular basis" (Imregi 2015). Also TV stations in Cox Media Group incorporated drones into their coverage (Mullin 2016).

According to the interviews conducted by Belair-Gagnon et al. (2017, p. 5) among early adopters of drone journalism in the US, the main reasons for using flying cameras were "low cost," "more precise visualization for storytelling," and "safe access to uncharted reporting terrains."

However, for three years, the development of American drone journalism was, to a large extent, put to rest due to flying restrictions by the government. The grip was loosened when the new FAA rules, more than 600 pages in length, came into effect in August 2016. The new national rules legalized drone journalism across the states. According to these drone regulations, users must be 16 years old, be able to understand English, and have an operator's certificate. In order to get the certificate one has to pass a knowledge test. Drones were not allowed to fly above 120 meters (FAA 2016; Belair-Gagnon et al. 2017). These simple rules align with international aviation regulations and are similar to the rules set forth in Europe.

In 2017, the American consensus emphasized that drones were already just another tool available to journalists. The newest models of camera drones, DJI Phantom 4s and Inspires of DJI Mavic were frequently used. Many TV stations were putting their drone operations in-house and wanted to focus on real-time broadcasting in breaking news situations. Interestingly, many stations were branding their drones. In Chicago, one TV station even named one of its drones and dedicated a webpage to it. Many restrictions still hindered the smooth use of drones in urban settings. Bad weather conditions were still enough to ground many drones (Perritt & Sprague 2017b).

Even after the new rules came into effect in the United States, authorities were able to limit the use of drones, especially during protest events, such as the case of the Dakota Access Pipeline demonstration in the fall of 2016. After the dramatic drone footage from Standing Rock (riot police blasting crowds with water cannons, rubber bullets and tear gas) went viral via social media platforms, the FAA banned all civilian use of drones within a four-mile radius of the area (Koebler 2016; Glaser 2016; Ahtone 2016; Kopstein 2016).

The Standing Rock case was, in reality, an imitation of the Ferguson case. As Joshua Kopstein (2016) writes: "Aerial images and video are often key to knowing crowd size and holding the police accountable for abuses against activists." The only difference was the media coverage by drones. Only after drone footage was repeatedly screened on television did mainstream media pour into Standing Rock. Also, according to Ahtone (2016):

> It's been entertaining to watch the press crowd come out to Indian Country. They didn't want to, of course, but after a few months of United States security forces using tear gas, rubber bullets, mace, water cannon and concussion grenades on hundreds of indigenous protesters intent on stopping an oil pipeline, they had to.

In Europe there were 14 illegal drone flights over French nuclear plants reported in 2014 alone. French authorities did not have any leads on who was behind the flights, but the police officers were under orders to shoot down any aircraft that could threaten the plants (Bilefskynov 2014). Also, a freelancer working for the BBC was arrested in 2014 while gathering footage related to a fatal fire near Gatwick airport in London (Quinn 2014).

Tension between journalists and authorities appears to be a common feature of the first wave of drone journalism. Simultaneously, it might be claimed that the first news imagery that came out of these tensions, controversies and even accidents helped to spread the word and increase awareness of the new flying cameras, and paved the way for the next cohort of drone journalists; these events actually catalyzed drone journalism to go mainstream.

Other seminal incidents

The first person the Federal Aviation Administration (FAA) tried to fine was someone who used a drone to film a commercial at the University of Virginia. The agency demanded a $10,000 fine for reckless flight of an aircraft. In the end, however, a federal judge ruled that it was legal to operate drones commercially (Koebler 2014). Later, in October 2015, the FAA fined a drone-photography company $1.9 million for allegedly conducting 65 drone flights in New York City and Chicago between March 2012 and December 2014 without the required authorization (Vanian 2015). In 2017, the drone-photography company agreed to pay a $200,000 penalty to settle allegations without acknowledging violating federal regulations (Jansen 2017).

The first Australian drone journalism controversy occurred when Channel Nine's 60 Minutes program used a drone to capture an aerial video and images of the Christmas Island immigration detention center after being denied entry to the facilities. At the end of its flying mission, the drone crashed into the Indian Ocean. No laws were broken, but the spokesman for the detention center blamed the drone journalists for causing fear and jeopardizing safety. Interestingly, Australia had already introduced the world's first drone legislation in 2002, but the speed of technological advances promptly made the old regulations obsolete (Corcoran 2012; Goldberg et al. 2013, p. 22).

In Australia, drones were used early for recording forest and bush fires. Such action caused some controversy. The Civil Aviation Safety Authority (CASA) detected two incidents where drone flights put firefighting responses at risk. The CASA warned that "flying a remotely piloted aircraft in the same airspace as helicopters and planes fighting fires 'creates a real risk of a mid-air collision.'" The authorities said that if they received evidence of drones being used in an unsafe manner, they would issue fines, probably amounting to many thousands of dollars (ABC.net.au 2013a; ABC.net.au 2013b). This controversy between hobbyists/journalists and fire fighters erupted later also in the US in 2015 (Fessenden 2015).

Drones made their first headlines in New Zealand when they crashed into buildings, such as a skyscraper in Auckland (Mortimer 2012b). In Finland, the first camera drone-related controversy started when local police threatened to shoot down all flying cameras used by journalists in a small plane's crash site in 2013. A freelancer took aerial images and videos of the crash scene and broadcast them during the national TV news. During the public debate, the freedom of the press advocates backed the use of drones for journalism and warned that shooting down any drones would be a crime (Lauk et al. 2016).

Sports, especially major leagues in various countries, have attracted drone journalists to test their skills. In Australia, National Rugby League and Twenty20 Big Bash League (cricket) were among the first leagues to use drone shots as promotional material (Corcoran 2013). In the US, hobbyists were the first people to use drones during sporting events. For example, according to the National Football League in 2014 at least 12 drones landed around stadiums on game days (Schmidt & Shearjan 2015). In the UK, a drone hobbyist was fined £1,800 by a court because of nine breaches of taking video over football grounds and tourist attractions in 2014 (BBC.com 2015a). In Finland, the biggest social media video hit in 2015 included drone footage. In the video, a Finnish world champion in orienteering runs up 426 stairs in a landfill trying to beat the one-minute record (Facebook 2015).

Later, in a World Cup slalom race at Madonna di Campiglio in Italy, a remotely controlled drone slammed down just inches behind a skier. After the incident, the International Ski Federation (FIS) stated that "an accident such as the drone crash cannot happen again" (Grez 2015). In Australia, a drone collided with an athlete during a triathlon competition, injuring the athlete (Grubb 2014); another crashed into a bridge and ended on train tracks (Cosier 2013). In addition, in the US, the FAA recorded hundreds of near-collisions between airplanes and drones (Whitlock 2015).

Many new safety features have already been developed because of these early incidents: geofencing offers one example. Geofencing is based on software, which automatically limits how high and how far you can fly your drone, and no-fly zones could also be programmed (CBS News/AP 2014; Grush 2016). In China, the first no-fly zone was set in 2013 at Tiananmen Square, Beijing. DJI, the Chinese manufacturer of small consumer and professional drones for aerial photography, placed a virtual fence around the city center in Beijing (Schroyer 2013b). Virtual fencing was also set for airports – the first places in China, and elsewhere – where hobbyists started to cause dangerous incidents with airplanes (Luo 2013).

Observers like Perritt and Sprague (2017b, p. 2) have argued that the barriers to the wider use of drones "are almost entirely political and regulatory, not technological." In our January 2017 interview with Ben Kreimer, one of the world's leading civilian drone consultants, Kreimer criticized countrywide bans on using drones:

> Bans, it is like you are saying we are not going to do anything about it. So it is not a way to move forward in terms of adopting technology into the society.

In conclusion, in spite of numerous controversies and even total bans in some countries, we propose that the first wave of drone journalism served as a creative outlet for the entrepreneurial potential of true news enthusiasm. While media managers hesitated because no short-term market advantages were in sight, drone journalist pioneers were driven by the weak ties of Internet networks and the new cultures of learning and risk taking by doing and sharing. And, we argue, the connected networks of the early camera drone pilots were what actually changed the games.

Our data indicate that it took less than a decade for camera drones to become a ubiquitous journalism tool in larger newsrooms of pioneering countries. As media users, we watch drone clips on most platforms without even noticing. Individuals who invested in the new flying robot are now in the competitive forefront. During the same period, drones used for journalism purposes have become advanced sensor platforms that challenge journalism in even new ways, legally as well as ethically. The infinite game of the technium is moving on.

References

ABC.net.au. 2013a. CASA warns bushfire drone operators of potential fines, www.abc.net.au/news/2013-10-27/casa-warns-bushfire-drone-operators-of-potential-fines/5048122 [Accessed 24 March 2017].

ABC.net.au. 2013b. Drone footage of fire damage at Lithgow, www.abc.net.au/news/2013-10-24/lithgow-drone-footage/5043670 [Accessed 24 March 2017].

Ahtone, Tristan. 2016. How media did and did not report on Standing Rock. *Vice.com*, https://news.vice.com/story/standing-rock-protests-what-comes-after [Accessed 2 April 2017].

BBC.news.com. 2013. "Hexacopter" flying camera makes BBC News debut, www.bbc.com/news/technology-24709180 [Accessed 24 March 2017].

BBC.com. 2015a. Man fined after flying drones over Premier League stadiums, www.bbc.com/news/uk-england-nottinghamshire-34256680, [Accessed 11 April 2017].

BBC.com. 2015b. Paris police arrest Al-Jazeera journalists over drone, www.bbc.com/news/world-europe-31632253 [Accessed 2 April 2017].

BBC.com. 2016. Migrant crisis: Drone captures life jacket "graveyard," www.bbc.com/news/world-europe-35867695 [Accessed 13 April 2017].

Belair-Gagnon, V., Owen, T., and Holton, A. E. 2017. Unmanned aerial vehicles and journalistic disruption. *Digital Journalism*. Published online 27 January. DOI:10.1080/21670811.2017.1279019.

Bilefskynov, D. 2014. France arrests 3 with drones by power plant. *Nytimes.com*, www.nytimes.com/2014/11/07/world/europe/3-found-with-drones-near-nuclear-plant-are-questioned-in-france.html [Accessed 11 April 2017].

Caputo, L. 2013. Drone Journalism, uno sguardo inedito sugli eventi. *Corrieredellasera.it*, http://piazzadigitale.corriere.it/2013/03/04/drone-journalism-uno-sguardo-inedito-sugli-eventi/ [Accessed 19 March 2017].

Castells, M. 2009. *Communication Power*. Oxford: Oxford University Press.

Castells, M. 2012. *Networks of Outrage and Hope: Social Movements in the Internet Age*. Chichester, UK: Wiley.

CBS News/AP. 2014. Drones' dangers confronted by Chris Anderson 3D robotics. *CBSnews.com*, www.cbsnews.com/news/drones-dangers-confronted-by-chris-anderson-of-3d-robotics/ [Accessed 28 March 2017].

Christensen, C. M. 1997. *The Innovator's Dilemma: When New Technologies Cause Great Firms to Fail*. Boston, MA: Harvard Business School Press.

Christoprudov, D. 2011. Kak mi nad Bolotnoi plosadju letali. [How we flew over Bolotnoi market place], http://chistoprudov.livejournal.com/87764.html.

Collins, K. 2014. Behind the mind-boggling shots captured by BBC drones. *Wired.co.uk*. www.wired.co.uk/article/bbc-drone-journalism [Accessed 9 January 2018].

Corcoran, M. 2012. Drone journalism takes off. *ABC.net.au*, www.abc.net.au/news/2012-02-21/drone-journalism-takes-off/3840616 [Accessed 18 March 2017].

Corcoran, M. 2013. Drones set for commercial takeoff. *ABC.net.au*, www.abc.net.au/news/2013-03-01/drones-set-for-large-scale-commercial-take-off/4546556 [Accessed 19 March 2017].

Corcoran, M. 2014. Drone journalism: Newsgathering applications on Unmanned Aerial Vehicles (UAVs) in covering conflict, civil unrest and disaster. *Cryptome.org*, https://cryptome.org/2014/03/drone-journalism.pdf [Accessed 10 May 2017].

Cosier, C. 2013. "I don't know whether it's a bomb or not": Train driver flummoxed after drone hits Sydney Harbour Bridge. *Smc.com.au*, www.smh.com.

au/technology/sci-tech/i-dont-know-whether-its-a-bomb-or-not-train-driver-flummoxed-after-drone-hits-sydney-harbour-bridge-20131126-2y76m.html [Accessed 26 March 2017].

Diep, F. 2014. Salvadoran newspaper sends drone to cover presidential election: *El drone* took photos and video of voters. *Popsci.com*, www.popsci.com/article/technology/salvadoran-newspaper-sends-drone-cover-presidential-election [Accessed 25 March 2017].

Dronejournalismlab.org. 2014. www.dronejournalismlab.org/page/7 [Accessed 25 March 2017].

DronesSkycam. 2013. *Youtube.com*, www.youtube.com/watch?time_continue=6&v=DLlD0tiA-80 [Accessed 9 January 2018].

Eriksson, K. 2017. Nu flyger det drönare i Sverige igen. *Dronesweden.se*, www.dronesweden.se/nyheter/nu-flyger-det-dronare-i-sverige-igen/ [Accessed 16 April 2017].

FAA. 2016. Operation and certification of small unmanned aircraft systems, www.faa.gov/uas/media/RIN_2120-AJ60_Clean_Signed.pdf, 2.4.2017.

Facebook. 2015. www.is.fi/muutlajit/art-2000000995577.html [Accessed 13 April 2017].

Fessenden, M. 2015. Drones are getting in the way of firefighters combating wilderness blazes. *Smithsonian.com*, www.smithsonianmag.com/smart-news/drones-are-preventing-firefighters-combating-wilderness-blazes-180956064/ [Accessed 11 April 2017].

Flanagan, B. 2015. Nepal quake: Media drone swarm shows "need for standards." *Alarabiya.net*, http://english.alarabiya.net/en/media/television-and-radio/2015/05/14/Nepal-quake-Media-drone-swarm-shows-need-for-standards-.html [Accessed 11 April 2017].

Glaser, A. 2016. The FAA banned drones from flying at the Standing Rock oil pipeline protest: Drone footage of clashes with police at the campsite have gone viral. *Recode.net*, www.recode.net/2016/11/28/13767216/faa-bans-drones-standing-rock-dakota-access-pipeline-video [Accessed 2 April 2017].

Goldberg, D., Corcoran, M., and Picard, R. G. 2013. Remotely Piloted Aircraft Systems and Journalism: Opportunities and challenges of drones in news gathering. https://ora.ox.ac.uk/objects/uuid:a868f952-814d-4bf3-8cfa-9d58da904ee3.

Greenwood, F. 2015. Thailand is cracking down on drones. *Slate.com*, www.slate.com/articles/technology/future_tense/2015/02/thailand_drone_regulations_why_you_should_care.html [Accessed 9 January 2018].

Grez, M. 2015. Drone crashes onto piste, misses champion skier by inches. *CNN.com*, http://edition.cnn.com/2015/12/23/sport/marcel-hirscher-drone-crash/ [Accessed 2 April 2017].

Grubb, B. 2014. "River of blood" after drone "hits" Australian athlete. *Smc.com.au*, www.smh.com.au/technology/technology-news/river-of-blood-after-drone-hits-australian-athlete-20140407-zqruh.html [Accessed 26 March 2017].

Grush, L. 2016. DJI just launched new software to stop its drones from flying in restricted airspace. *Theverge.com*, www.theverge.com/2016/1/1/10699078/dji-drones-geofencing-software-restrictions-flying [Accessed 13 April 2017].

Gynnild, A. 2013. The robot eye witness: Extending visual journalism through drone surveillance. *Digital Journalism*, Published online 20 February 2014, 334–343. http://dx.doi.org/10.1080/21670811.2014.883184.

Halliday, J. 2015. Three BBC journalists questioned for using drone in Davos no fly zone. *Theguardian.com*, www.theguardian.com/media/2015/feb/02/three-bbc-journalists-questioned-drone-no-fly-zone-davos [Accessed 2 April 2017].

Haner, J. 2015. A drone's vantage point of a melting Greenland. *Nytimes.com*, www.nytimes.com/2015/10/28/insider/a-drones-vantage-point-of-a-melting-greenland.html?_r=0 [Accessed 12 April 2017].

Hill, K. 2011. FAA looks into news corp's daily drone, raising questions about who gets to fly drones in the U.S. *Forbes.com*, www.forbes.com/sites/kashmirhill/2011/08/02/faa-looks-into-news-corps-daily-drone-raising-questions-about-who-gets-to-fly-drones-in-the-u-s/#3e32492533b1 [Accessed 12 March 2017].

Holton, A. E., Lawson, S., and Love, C. 2015. Unmanned aerial vehicles: Opportunities, barriers, and the future of "drone journalism." *Journalism Practice*, Vol. 9, No. 5, 634–650.

Imregi, L. 2015. Drones: A unique view on storytelling. *Ap.org*, https://insights.ap.org/whats-new/drones-a-unique-view-on-storytelling [Accessed 12 April 2017].

Jakarta P. 2015. Govt rule limits drone use in journalism. *Thejakartapost.com*, www.thejakartapost.com/news/2015/09/22/govt-rule-limits-drone-use-journalism.html [Accessed 11 April 2017].

Jansen, B. 2017. Drone-photography company fined $200,000 by FAA. *Usatoday.com*, www.usatoday.com/story/news/2017/01/17/faa-drone-skypan/96671342/ [Accessed 16 May 2017].

Johnson, E. 2015. Kenya basically bans all drone use: Despite potential benefits they may yield. *Pri.org*, www.pri.org/stories/2015-12-15/kenya-basically-bans-all-drone-use-despite-potential-benefits-they-may-yield [Accessed 12 April 2017].

Kelly, K. 2010. *What Technology Wants*. New York: Viking.

Keneally, M. 2012. Hunters take aim at an animal rights group's video drone. *Dailymail.co.uk*, www.dailymail.co.uk/news/article-2103803/Hunters-aim-animal-rights-groups-video-drone.html [Accessed 18 March 2017].

Koebler, J. 2014. Commercial drones are completely legal, a federal judge ruled. *Vice.com*, https://motherboard.vice.com/en_us/article/commercial-drones-are-completely-legal-a-federal-judge-ruled [Accessed 26 March 2017].

Koebler, J. 2016. The government is using a no fly zone to suppress journalism at Standing Rock. *Vice.com*, http://motherboard.vice.com/read/the-government-is-using-a-no-fly-zone-to-suppress-journalism-at-standing-rock [Accessed 2 April 2017].

Kopstein, J. 2016. Police are making it impossible to use drones to document protests, www.vocativ.com/396662/police-drone-journalists-protests/ [Accessed 2 April 2017].

Lam, B. 2014. Incredible close-up drone video of an erupting volcano in Iceland. *Wired.com*, www.wired.com/2014/10/drone-video-iceland-eruption-bardarbunga-volcano [Accessed 28 March 2017].

Lauk, E., Uskali, T., Kuutti, H., and Hirvinen, H. 2016. Drone journalism: The newest global test of press freedom. In Carlsson, Ulla (Ed.), *Freedom of Expression and Media in Transition: Studies and Reflections in the Digital Age*, pp. 117–125. Gothenburg: Nordicom.

Lowy, J. 2013. FAA warns public against shooting guns at drones. *AP.org*, www.denverpost.com/2013/07/19/faa-warns-public-against-shooting-guns-at-drones/ [Accessed 19 March 2017].

Luo, C. 2013. Police arrest four in Beijing after mysterious drone forces diversion of civilian planes. *South China Morning Post, Scmp.com*, www.scmp.com/news/china-insider/article/1393833/police-arrest-four-after-mysterious-drone-forces-diversion [Accessed 8 April 2017].

Mackley, G. 2012. Most incredible volcano expedition ever 2012: The full version. *Youtube.com*, www.youtube.com/watch?v=VuQrUwFn6bU [Accessed 8 April 2017].

Martinelli, N. 2011. Five things you need to know about drone journalism. *Ijnet.org*, http://ijnet.org/en/blog/five-things-you-need-know-about-drone-journalism [Accessed 12 March 2017].

Mortimer, G. 2012a. Dallas meat packing plant investigated after drone images reveal pollution. *Suasnews.com*, www.suasnews.com/2012/01/dallas-meat-packing-plant-investigated-after-drone-images-reveal-pollution/ [Accessed 18 March 2017].

Mortimer, G. 2012b. Multirotor hits skyscraper and burns, Auckland, NZ. *Suasnews.com*, www.suasnews.com/2012/10/multirotor-hits-skyscraper-and-burns-downtown-auckland-nz/ [Accessed 5 April 2017].

Mullin, B. 2016. Why 2016 could be a break out year for drone journalism. www.poynter.org/news/why-2016-could-be-breakout-year-drone-journalism [Accessed 9 January 2018].

NBCnews.com. 2015. Vanuatu locals describe "hell on earth" after cyclone pam, www.nbcnews.com/nightly-news/video/vanuatu-locals-describe–hell-on-earth–after-cyclone-pam-413998659736 [Accessed 2 April 2017].

Pead, S. 2014. Crucial time for drone reporting laws: Industry needs to act. *Wan/Ifra.org*, https://blog.wan-ifra.org/2014/06/27/crucial-time-for-drone-reporting-laws-industry-needs-to-act [Accessed 26 March 2017].

Penhaul, K. 2013. A bird's eye view of Haiyan devastation. *CNN.com*, http://edition.cnn.com/videos/world/2013/11/18/philippines-drone-camera-penhaul.cnn [Accessed 25 March 2017].

Perritt, H. H. Junior, and Sprague, E. O. 2017a. *Domesticating Drones: The Technology, Law, and Economics of Unmanned Aircraft*. London: Routledge.

Perritt, H. H. Junior, and Sprague, E. O. 2017b. Navigating cautiously: Tentative drone journalism. *Rtdna.org*, www.rtdna.org/article/navigating_cautiously_tentative_drone_journalism [Accessed 8 May 2017].

Pitt, F. 2014. *Sensors and Journalism: Tow Center for Digital Journalism*. New York: Columbia Journalism School.

Postema, S. 2015. News drones: An auxillary perspective: Promising applications of the news drone as a supplementary tool for news gathering, analyzing and publishing. MA thesis. Edinburgh Napier University, Edinburgh Napier.

Quinn, B. 2014. Photojournalist arrested after filming with drone near Gatwick airport. *Theguardian.com*, www.theguardian.com/uk-news/2014/dec/31/drone-photojournalist-arrested-gatwick-aiport-near [Accessed 2 April 2017].

Rogers, E. M. 1995. *Diffusion of Innovations*. Third Edition. New York: The Free Press.

Schmidt, M. S., and Shearjan, M. D. 2015. Drones spotted, but not halted, raise concerns. *Nytimes.com*, www.nytimes.com/2015/01/30/us/for-super-bowl-and-big-games-drone-flyovers-are-rising-concern.html [Accessed 11 April 2017].

Schroyer, M. 2012. Update on hunters shooting down activist drone: On-board footage, lawsuits and more. *Mentalmunition.com*, www.mentalmunition.com/2012/02/update-on-hunters-shooting-down.html [Accessed 18 March 2017].
Schroyer, M. 2013a. BBC touts benefits of its new hexacopter drone. *Dronejournalism.org*, www.dronejournalism.org/news/2013/10/bbc-touts-benefits-of-its-new-hexacopter-drone [Accessed 8 April 2017].
Schroyer, M. 2013b. DJI blocks flying above Tiananmen Square with software update. *Dronejournalism.org*, www.dronejournalism.org/news/2013/8/dji-blocks-flying-above-tiananmen-square-with-software-update [Accessed 5 April 2017].
Schroyer, M. 2014. Q&A with the photographer who explored Chernobyl with a drone. *Dronejournalism.org*, www.dronejournalism.org/news/2014/8/qa-with-the-photographer-who-explored-chernobyl-with-a-drone [Accessed 8 April 2017].
Schroyer, M. 2015. PSDJ creates 3D models and maps of war-ravaged Donetsk Airport, using citizen groups' drone video. *Dronejournalism.org*, www.dronejournalism.org/news/2015/1/psdj-creates-3d-models-and-maps-of-war-ravaged-donetsk-airport-using-citizen-groups-drone-video [Accessed 8 April 2017].
Shammas, J. 2015. Nepal earthquake: Drone captures incredible footage over Kathmandu showing aftermath of disaster. *Mirror.co.uk*, www.mirror.co.uk/news/world-news/nepal-earthquake-drone-captures-incredible-5592151 [Accessed 11 April 2017].
Siemens, G. 2005. Connectivism: A learning theory for the digital age. www.itdl.org/journal/jan_05/article01.htm.
Sky.com. 2015. Aerial footage shows devastation of Nepal earthquake, http://news.sky.com/video/aerial-footage-shows-devastation-of-nepal-earthquake-10361825 [Accessed 11 April 2017].
Teirstein, Z. 2016. Sweden places ban on flying camera drones without surveillance permits. *Theverge.com*, www.theverge.com/2016/10/24/13381050/sweden-bans-surveillance-camera-drones-permits [Accessed 16 April 2017].
Theguardian.com. 2015. Drone footage shows scale of destruction in Donetsk, Ukraine: Video, www.theguardian.com/world/video/2015/jan/17/drone-destruction-donetsk-ukraine-airport-video [Accessed 11 April 2017].
Theguardian.com. 2016. Drone footage of homes in Syria shows utter devastation: Video, www.theguardian.com/world/video/2016/feb/04/drone-footage-homs-syria-utter-devastation-video [Accessed 13 April 2017].
Thelocal.se. 2016. Sweden set to reverse controversial "drone ban," www.thelocal.se/20161220/sweden-set-to-reverse-controversial-drone-ban [Accessed 16 April 2017].
Thetandd.com. 2012. Animal rights group says drone shot down, http://thetandd.com/animal-rights-group-says-drone-shot-down/article_017a720a-56ce-11e1-afc4-001871e3ce6c.html [Accessed 18 March 2017].
Thomas, D. and Brown, J. S. 2011. *A New Culture of Learning: Cultivating the Imagination for a World of Constant Change*. Toronto: Podium Publishing.
Tremayne, M., and Clark, A. 2013. New perspectives from the sky: Unmanned aerial vehicles and journalism. *Digital Journalism*, Vol. 2, No. 2, 232–246.
Uskali, T., and Lauk, E. 2018. Keeping reporters safe: Ethics of drone journalism in a humanitarian crisis. In Andersen, Robin, and de Silva, Purnaka L. (Eds.), *Routledge Companion to Media and Humanitarian Action*. London: Routledge.

Vanian, J. 2015. Drone startup issued biggest fine ever for flying without permission. *Fortune.com*, http://fortune.com/2015/10/06/faa-drone-fine-skypan/ [Accessed 12 April 2017].

Vaux, P. 2015. Drones find Russian base inside Ukraine. *Thedailybeast.com*, www.thedailybeast.com/articles/2015/06/30/apparent-russian-base-found-in-ukraine.html?source=TDB&via=FB_Page [Accessed 11 April 2017].

Walsh, N. P. 2015. Syrian town tries to rise from ashes after ISIS defeat. *Cnn.com*, http://edition.cnn.com/2015/05/05/middleeast/syria-kobani-rising-from-ashes/ [Accessed 11 April 2017].

WAN/Ifra. 2015. Al Jazeera arrests in Paris a reminder to drone responsibly. *Blog.wan-ifra.org*, https://blog.wan-ifra.org/2015/02/26/al-jazeera-arrests-in-paris-a-reminder-to-drone-responsibly [Accessed 2 April 2017].

Westcott, R. 2013. "Hexacopter" changes the way TV reporters work. *BBC.com*, www.bbc.com/news/business-24712136 [Accessed 29 October 2017].

Whitlock, C. 2015. FAA records detail hundreds of close calls between airplanes and drones. *Washingtonpost.com*, www.washingtonpost.com/world/national-security/faa-records-detail-hundreds-of-close-calls-between-airplanes-and-drones/2015/08/20/5ef812ae-4737-11e5-846d-02792f854297_story.html?utm_term=.084b0b599a25 [Accessed 11 April 2017].

Wilonsky, R. 2013. Columbia Packing Company, known for dumping pig blood into Trinity River, gets OK from city of Dallas to reopen. *Dallasnews.com*, www.dallasnews.com/news/dallas-city-hall/2013/06/28/columbia-packing-company-gets-ok-from-city-of-dallas-to-reopen-as-meatpacker-not-slaughterhouse [Accessed 18 March 2017].

Wu, T. 2011. *The Master Switch: The Rise and Fall of Information Empires*. New York: Vintage Books.

Zavrsnik, A. (ed.). 2016. *Drones and Unmanned Aerial Systems: Legal and Social Implications for Security and Surveillance*. Berlin: Springer.

3 Dronalism, newsgathering protection and day-to-day norms

David Goldberg

Introduction

This chapter makes three claims; 1) The standard tripartite division of drone, operator and drone applications is *conceptually* inadequate. 2) Using drones for journalism newsgathering – "dronalism" – is unique among drone applications because it is a constitutionally or rights-protected activity. 3) For operators – *a fortiori* including those using drones for journalism – the most significant day-to-day issue is knowing how their activity will be policed, including the possibility of prosecution and involvement with courts, especially concerning the issue of "low flying."

However, using drones for journalism is only *relatively* disruptive. Given that such a use for a drone is to turn the drone into a newsgathering flying camera, one should remember the earlier innovative disruption created by John Silva in 1958. He converted a small helicopter into the first airborne virtual television studio. The KTLA "Telecopter," as it was called by the Los Angeles station where Silva was the chief engineer, became the basic tool of live television traffic reporting, disaster coverage and that most famous glued-to-the-tube moment in the modern era of celebrity gawking, the 1994 broadcast of O. J. Simpson leading a motorcade of pursuers on Los Angeles freeways after his former wife and a friend of hers were killed (Pool 2012).

The concept of drone journalism was first explored in 2002 at The Poynter Institute for Media Studies by Larry Larsen, who looked at the ethical and practical uses of unmanned aerial vehicles for reporting and research. More recently, the topic has been considered by the Reuters Institute at the University of Oxford (Goldberg, Corcoran and Picard 2013)

In general, the standard highest-level division is between non-civilian and civilian operators and purposes. Non-civilian includes military and government, including law enforcement. The next level of distinction is between commercial and non-commercial (i.e., recreational, operators and purposes). But this schema ignores or elides the use of drones for journalism newsgathering, which does not fit any of the three categories, being *sui*

DOI: 10.4324/9781315163659-3

generis. The conceptual point was made in the Brief of news media *amici* in support of Raphael Pirker (2014).

In criticizing the Federal Aviation Administration's (FAA) ban (at the time), the Brief urges that the FAA's position is untenable, as it rests on a *fundamental misunderstanding* about journalism. Newsgathering is not a "business [aka commercial] purpose . . . journalism is not like other businesses". The Brief points to the (legal) fact that, in the US, the Supreme Court recognizes that the publication of news is not a "commercial" activity comparable to the sale of goods and services (Brief of news media *amici* in support of Raphael Pirker 2014):

And, memorably, in *Murdock v. Com. of Pennsylvania*, the Court stated:

> It should be remembered that the pamphlets of Thomas Paine were not distributed free of charge.
>
> (*Murdock v Pennsylvania* 319 U.S. 105, 110 (1943))

Dronarazzi

Using a drone to capture images may look like but it does not *per se* make the activity dronalism, or responsible newsgathering. The latter can be easily distinguished from activities that *prima facie* seem analogous but are clearly distinguishable (e.g., are simply voyeurism). There is almost a willful refusal to accept the perfectly reasonable proposition advanced by the Australian Law Commission (ALRC 2013) that proper journalism can be meaningfully distinguished from:

> circumstances which are not journalistic in nature, where the public interest in a matter is trivial, or where the matter is merely of interest to the public or for the purposes of gossip.

Further, as the Commission stated:

> Surveillance will sometimes be *necessary and justified* when conducted in the course of responsible journalistic activities. . . . Media and journalistic activities offer significant public benefit, and these activities may at times justify the use of *surveillance devices* without the notice or consent of the individuals placed under surveillance.
>
> (emphases added)

The key claim of this section is that deploying drones in the context of dronalism engages a basic human right, namely, the right of the public to receive ideas and information. This includes the component right(s) needed to make it a reality, the whole being part of the general right to freedom of expression. Thus, the use of drones by media companies and/or citizen journalists and, crucially, any restrictions thereon, particularly by blanket bans, raise unique concerns because such use engages elements of the right to freedom of expression. Actually, the threshold right is, arguably, the right to access the communications technology that a drone is in such situations (Herr 2013). By implication, any general restriction(s) on using drones which, *a fortiori*, includes newsgathering, would amount to a *prima facie* infringement of this right. Only if, *in casu*, extremely strong compelling overriding considerations defending and promoting another protected interest, would the right to use a drone in that specific, fact-limited context be trumped. At the very least, *in the absence of carrying out an explicit exercise balancing the competing interest involved*, any restriction would be challengeable as procedurally flawed. (Goldberg 2015.)

The "balancing" requirement was well expressed (in the context of privacy v. freedom of expression) by Ofcom, the UK media regulator:

> In Ofcom's view, the individual's right to privacy has to be balanced against the competing right of the broadcaster to freedom of expression. Neither right as such has precedence over the other and where there is a conflict between the two, it is necessary to intensely focus on the comparative importance of the specific rights. Any justification for interfering with or restricting each right must be taken into account and any interference or restriction must be proportionate.
>
> (Ofcom 2017)

The bottom line is, as the U.S. Supreme Court stated in Branzburg v Hayes (1972):

> without some protection for *seeking out* the news, freedom of the press could be eviscerated.
>
> (emphasis added)

Newsgathering and news reporting are strongly protected by US law, including the First Amendment to the Constitution. The public relies on an independent press to gather and report the news and ensure an informed public. For this reason, the National Telecommunications and Information Administration's (NTIA) 2016 publication, *Voluntary Best Practices for UAS Privacy, Transparency, and Accountability*, accepted that "these Best Practices do not apply to newsgatherers and news reporting organizations." Newsgatherers and news reporting organizations may use an unmanned aircraft system (UAS) in the same manner as any other comparable technology to capture, store, retain and use data or images in public spaces.

Newsgatherers and news reporting organizations should operate under the ethics rules and standards of their organization, and according to existing federal and state laws. (Ntia.doc.gov 2016).

The most significant day-to-day concern

The most significant day-to-day concern for operators is knowing how their activity will be policed, including the likelihood of being prosecuted and dealt with by the courts, notably concerning the issue of "low flying."

It is difficult enough to know and keep track of the changes concerning drone laws/rules in any country, region and globally (Martin 2017; Patterson 2016; Storyhunter.com 2016). But this section raises the question and issue: what are the sources for knowing this level of knowledge affecting the day-to-day operations and are they easily publicly accessible? The situation, at least in the UK, is that operators have access to – and would be well advised to read – the *Memorandum of Understanding* between the National Police Chiefs Council, the Civil Aviation Authority, Home Office and Department of Transport, signed to clarify roles of each in relation to investigation and prosecution of drone offenses as set out in the Air Navigation Order 2009 (as revised) (MOU). Additionally, this document should be on the bedside table of every dronalism operator: the National Police Chiefs Council's *Guidance to [Police] Officers on DRONES: Legislation and Dealing with Misuse* (NPCC 2017).

The extent of such enforcement intervention is a matter of dispute. For the UK, the numbers of complaints to the police do seem to be growing – to 3,456 in 2016 compared with 283 in 2014. The 2016 figure was almost three times higher than the 2015 total of 1,237 incidents. Complaints include allegations of snooping, burglary "scoping" exercises, mid-air near-misses and the smuggling of contraband into prisons (Siddique 2017). However, not unreasonably, in 2016, the DJI company (Dà-Jiāng Innovations) rather tartly put out a press release noting that the report about numbers of complaints:

> is simply an unedited listing of raw complaints, with no attempt to verify whether any of them had any merit. As we saw with the recent incident when a drone that supposedly hit a plane was really a plastic bag, initial reports cannot be taken at face value.
>
> (DJI.com 2016)

So far, there have not been, in the UK, any court-imposed sentences for dronalism. The most severe disposals – resulting in custodial sentences – have involved smuggling drugs, mobile phones and other items into prisons using a drone. At least four men are serving sentences ranging from 14 months to over five years.

There have been a number of legal challenges to using drones in the context of journalism or newsgathering. Most notably, the Dutch Journalists Association and Rene Oudshoorn took their government before the Hague District Court complaining that the difference between the then-existing rules covering journalists (professionals) and recreational users was a breach of the formers' European Court of Human Rights Article 10 rights. In an interview with the present author, Oudshoorn said that: "the court decided that there were no grounds for our complaint due to new regulations that give some more freedom to drone journalism. The amateur still has much more freedom but that's going to change over time."

In the US, Pedro Rivera was:

> operating a small unmanned aircraft system, a DJI Phantom 2 Vision, near the scene of a fatal car accident . . . when he was questioned by police. WFSB, the television station where Rivera was employed, suspended the photographer for a week without pay after being contacted by police through email and phone calls.
>
> (Dronejournalism.org 2014)

More recently, arising out of coverage of the Dakota Access Pipeline dispute, Myron Dewey, owner of Digital Smoke Signals, a website that includes news, videos and forums, was charged with stalking for allegedly using a drone to film private security guards (Dalrymple 2016). However, subsequently, the criminal charges were dropped.

Anecdotally, Greg Agvent, Senior Director, CNN Aerial Imagery & Reporting, told the present author (April 3, 2017) that CNN had experienced hardly any legal "pushback" against using drones for newsgathering, as, "principally we don't assume we have a right to fly – we don't operate recklessly. We communicate our intentions in advance – we work with local, state and federal authorities when necessary – rule of thumb, you can't over communicate."

What are some of the main – in this context this amounts to the most frequently mentioned – legal and regulatory issues relating to drones? The standard list includes safety; liability; insurance cover; certification and training of equipment and users; licensing and permission to fly; frequency spectrum allocation, assignment and license; lawful and unlawful interception of communications and registration. With regard to dronalism, respect for private life; trespass; nuisance and the right to photograph should be added. Two further topics should be mentioned: (i) whether the media may use pictures, video or other information collected by a third party using

UAS, and (ii) whether a person who sells images collected by UAS requires authorization for his or her operations (FAA 2015).

Drone usage in low-altitude airspace

This chapter, however, raises the less-often noted, specific and very long-standing issue of ownership of airspace (Klein 1959). Crucially, in the context of dronalism (indeed, drones generally), the narrower specific issue is about "low altitude airspace." Paul Voss opts for a zoned model for airspace whereby "UAS would only be allowed onto private property (below 300–500 feet altitude) with permission from the landowner, a warrant from a judge, or in the event of a public emergency" (Voss 2013). The assumption here is that dronalism operators would generally opt or prefer to function in the zone known as the "sweet spot," (i.e., lower than a helicopter and higher than a crane). But as Troy Rule observes:

> Regrettably, legal academicians and policymakers have devoted far less attention to an unsettled property law question that underlies these and many other domestic drone issues: Up to what height do surface owners hold strict rights to exclude flying objects from physically invading the airspace above their land? Legal uncertainty and confusion are likely to continue swirling around the domestic drone industry until courts or legislators clear up this basic property question. . . . Prior to the advent of modern drones, there was no pressing need to precisely define the scope of landowners' property interests in low altitude airspace. Unfortunately, as a growing flock of domestic drones stands ready for takeoff, *ambiguous* airspace rights laws are now threatening to impede the growth of an important new industry.
>
> (Rule 2015; emphasis added)

Rule advocates for new laws:

> expressly entitling landowners to exclude drones from the airspace above the surface of their land to a height of 500 feet in most locations. Such laws would at last provide a definite ceiling to the three-dimensional column of space initially allocated to surface owners under the common law's *ad coelom* doctrine. By establishing clearer entitlements in low-altitude airspace and creating a solid legal backdrop from which to layer supplemental rules, these laws would be a valuable step toward the more orderly and efficient integration of drone technologies [in the United States].
>
> (Rule 2015)

This chapter does not come to a concrete determination as to the precise altitude in feet or meters above the solum that the lower airspace shades or transitions into "navigable airspace," or indeed, the basis in common law, statute or regulatory fiat. However, one answer might be that a landowner can claim up to at least 83 feet – the height above Mr. Causby's chickens US Air Force planes flew to land (*United States v. Causby*). In 2016, California lawmakers approved a bill for a drone no-fly zone up to 350 feet above private property – but Governor Brown vetoed the bill.

But not coming to such a determination is not shying away from the issue because, as will be dealt with more fully, *infra*, even if the owner has ownership rights in lower/near ground airspace, *the right will not be absolute*. And in the context of responsible journalism, there would be a *defense*to any alleged infringement because of the *balancing* exercise that is involved when rights collide: here the right to enjoy – or exclude from access – one's property v. right to receive information.

In any case, one matter which is really important legally but hardly considered or even mentioned is *evidencing* the height and position of the drone relative to a property. Who decides and on the basis of what evidence? This was brought out starkly in the dispute between the so-called "drone slayer," William Merideth, and John David Boggs:

> The two men disagree how low Boggs flew his drone above Merideth's home – Merideth estimates about 100 feet or less, while Boggs has data that places it above 200 feet. The drone's exact altitude may not seem crucial, but it is unclear if landowners get to *decide* who can fly a drone over their property at 100 or even 300 feet.
>
> (Sneed 2015)

In an action in England, Bernstein claimed the aircraft (not a drone) was photographing over his property, whereas:

> Skyviews, who admitted taking the photograph, denied entering the airspace to do so. They claimed it was taken while the aircraft was flying over adjoining land. Skyviews, who instructed their pilots to photograph buildings, the owners of which might prove likely customers, claimed that their photograph was taken at a height of some 630 ft and 30 metres outside the boundary of Coppings Farm. . . . However, on any view of the evidence the plane would have flown close to the border of Coppings Farm at the time the photograph was taken and the probabilities were that at some time it had flown over the land.
>
> (*Bernstein v. Skyviews and General 1978*)

Another legally significant matter is *the extent and degree* of the "invasion" of a property owner's lower airspace as necessary for a finding of a *legal* cause of action. Thus, in Causby, the Court referred to the:

> Flights over private land are not a taking, *unless they are so low and so frequent* as to be a direct and immediate interference with the enjoyment and use of the land.
>
> (emphasis added)

In Bernstein, having dismissed the claim that the overflight constituted a trespass, the Court said:

> At the same time, however, the present judgment should not be understood as deciding that in no circumstances could a successful action be brought against an aerial photographer to restrain his activities. The judgment was far from saying that *if a plaintiff was subjected to the harassment of constant surveillance of his house from the air, accompanied by the photographing of his every activity*, the courts would not regard such a monstrous invasion of his privacy as an actionable nuisance for which they would give relief.
>
> (emphasis added)

These nuanced judgments qualify what are otherwise usually bald and simplistic binary assertions of right versus wrong when it comes to low-flying drones.

It is important to distinguish descriptive and prescriptive accounts of the matter (i.e., do property owners have an ownership right to lower/near ground airspace and, if so, what should the upper limit be before reaching currently accepted "navigable airspace"?) The thrust of this chapter is to urge the reader to recalibrate his or her thinking. Legally what matters is whether there is a legal defense open to the dronalist? So, while there might well be an infringement of another's right, the defense makes the infringement *justifiable*. In a non-aviation setting, the point is exemplified by the finding by the European Court of Human Rights in *Haldimann and Others v. Switzerland*, which balanced someone's right to respect for their privacy against the public interest in the receipt of information:

> a decision of the European Court of Human Rights (the "ECtHR") published on 24 February 2015, backed the investigative methods of four Swiss journalists who had used *hidden cameras* to expose the malpractice of insurance brokers. The ECtHR found by a majority decision that the journalists' criminal conviction by the domestic courts and an order to pay a number of small fines violated their right to freedom of expression as

> guaranteed by Article 10 of the European Convention of Human Rights. It was the first time the ECtHR examined the use of hidden cameras by journalists in a case where the person filmed was targeted as a representative of a particular profession rather than in a personal capacity.
>
> (emphasis added)

Extrapolating to dronalism, this decision and its reasoning offers a counterbalance to the blanket criticism of the use of small drones at very low altitudes which at first blush may seem to be a nuisance, or a trespass or an infringement of someone's right to respect for their privacy. For if the purpose of the use of the drone is to obtain video or audio evidence regarding a matter of general public interest or debate and is done in a limited and temporary way, *then such considerations can outweigh the infringement as being justifiable in the circumstances.*

A snapshot of the challenge facing dronalism is the advocacy call made by the American Society of Media Photographers (ASMP 2017). The ASMP is attempting to persuade committee members of the Texas Senate Committee on Business and Commerce to vote for an exemption from the 2013 law that currently makes drone photography illegal in many circumstances. The Bill, SB 839, will add "newsgathering" to the list of activities that are exempt from the law, therefore making it legal to use drones for journalistic activity.

More specifically, this chapter also suggests that the 2014 UK House of Lord's inquiry suggestion be utilized. In 2014, the UK House of Lords held an Inquiry into *Civilian Use of Drones in the EU* (Civilian Use of Drones in the EU, 2014). It considered "the use of RPAS (Remotely Piloted Aircraft Systems) by the media in order to capture images and videos." An interesting proposition was made by a witness that "authorities should consider recommending a data protection and airspace permission exemption for rapid response RPAS journalism. . . . If this particular developing area of rapid response journalism by RPAS is ignored then irresponsible, amateur cameramen will, in all likelihood, attempt to take footage anyway." The Committee concluded: "While journalists can use RPAS to enhance the reporting of important events, they can also be used to invade people's privacy. *UK media regulators should initiate a public consultation on the appropriate use of RPAS by the media, with a view to providing clear guidance.*" (Emphasis added). So far, the Recommendation has not been acted upon and awaits implementation, which in the present author's opinion, should be supported.

Despite the evidentiary problems mentioned, it seems useful for there to be at least nationally a height above ground level which would constitute the zone of lower/near ground airspace. It could be instituted for a period and be subject to review. However, as emphasized already, this would not really affect dronalism, as this use for a drone is a protected activity and permits the dronalist to infringe the otherwise protected right of airspace ownership.

Finally, there are a number of attempts to document the increasing plethora of national and regional norms affecting drone usage, including dronalism. A notable newcomer at the time of writing are the surveys published by the Center for the Study of the Drone at Bard College (Arthur Holland Michel and Dan Gettinger 2017). However, these only cover the US. One Recommendation of this chapter is that the surveys be conducted to cover, at least, European jurisdictions – and in particular the one covering drone incidents, since this would reflect this chapter's view that what is crucial for dronalism and dronalists (indeed, for all drone operators) is how and on what basis day-to-day incidents are handled and disposed of by the police and courts.

References

ALRC. (2017) Surveillance Devices: Responsible Journalism and the Public Interest. [online]. Available at: www.alrc.gov.au/publications/14-surveillance-devices/responsible-journalism-and-public-interest [Accessed 9 June 2017].

ASMP. (2017). Advocacy Alert: Support Needed for Texas Drone Journalism Bill SB839. www.asmp.org/advocacy/advocacy-alert-support-needed-texas-drone-journalism-bill-sb839/, April 2017.

Bernstein of Leigh v Skyviews & General Ltd. (1978) QB 479.

Branzburg v. Hayes. (1972) 408 U.S. 665, 702.

Dalrymple, A. (2016) Two Face Charges for Operating Drones during Pipeline Protests. *Droning On*, [online]. Available at: https://droningon.areavoices.com/2016/10/31/2-face-charges-for-operating-drones-during-pipeline-protests/ [Accessed 9 June 2017].

DJI.com. (2016) DJI Statement on Drone Complaints in UK. [online]. Available at: www.dji.com/newsroom/news/dji-statement-on-drone-complaints-in-uk- [Accessed 9 June 2017].

Dronejournalism.org (2014). Drone lawyer: First Amendment right to photograph extends to drone journalism. www.dronejournalism.org/news/2014/2/drone-lawyer-first-amendment-right-to-photograph-extends-to-drone-journalism.

FAA. (2015). Media Use of UAS, FAA May 5, 2015, www.scribd.com/doc/264441605/FAA-Memo-on-Media-Use-of-FAA

Goldberg, D. M. (2015) Dronalism: Journalism, Remotely Piloted Aircraft, Law and Regulation. *FIU Law Review*, [online], 10(2). Available at: http://ecollections.law.fiu.edu/cgi/viewcontent.cgi?article=1255&context=lawreview [Accessed 12 September 2017].

Goldberg, D. M. Corcoran and R. Picard. (2013) Remotely Piloted Aircraft Systems & Journalism Opportunities and Challenges of Drones in News Gathering. Available at: https://reutersinstitute.politics.ox.ac.uk/our-research/remotely-piloted-aircraft-systems-and-journalism [Accessed 9 June 2017].

Herr, R. E. (2013) Can Human Rights Law Support Access to Communication Technology? *Information & Communications Technology Law*, [online], 22(1). Doi: www.tandfonline.com/doi/abs/10.1080/13600834.2013.774517

Holland. M.A. & Gettinger, D. (2017) 'Drone Incidents: A Survey of Legal Cases', http://dronecenter.bard.edu/drone-incidents/, April 2017

Klein, H. (1959) Cujus est solum ejus est . . . quousque tandem? Journal of Air Law and Commerce, 26: 237–254.

Martin, J. (2017) Where to Fly a Drone in the UK and Abroad. *Pcadvisor.co.uk/TechAdvisor*. Available at: www.pcadvisor.co.uk/feature/gadget/where-fly-drone-in-uk-abroad-3620507/ [Accessed 9 June 2017].

National Telecommunications & Information Administration. (2016) Voluntary Best Practices for UAS Privacy, Transparency, and Accountability: Consensus, Stakeholder-Drafted Best Practices Created in the NTIA-Convened Multistakeholder Process. Available at: www.ntia.doc.gov/files/ntia/publications/uas_privacy_best_practices_6-21-16.pdf [Accessed 9 June 2017].

NPCC National Police Chiefs' Council. (2017) Guidance to Officers on Drones: Legislation and Dealing with Misuse. Available at: www.npcc.police.uk/Publication/NPCC%20FOI/Operations/175%2015%20NPCC%20Response%20Att%2001%20of%2001%2026102015.pdf [Accessed 9 June 2017].

Ofcom. (2017) Ofcom Broadcasting and Demand on Bulletin. [online], 327. Available at: www.ofcom.org.uk/__data/assets/pdf_file/0013/101227/Issue-327-of-Ofcoms-Broadcast-and-On-Demand-Bulletin.pdf, p. 91 [Accessed 13 September 2017].

Oudshoorn, R. (2017) *Telephone Interview*.

Patterson, J. (2016) Heliguy's Guide to Global Drone Regulations. *Heliguy*, [online]. Available at: www.heliguy.com/blog/2016/06/14/heliguys-guide-to-global-drone-regulations/#respond [Accessed 9 June 2017].

Pirker, R. (2014). Brief of News Media *Amici* in Support of Respondent Raphael Pirker. 2014. www.hklaw.com/files/Uploads/Documents/CaseBriefs/MediaLaw/Drones.pdf.

Pool, B. (2012) John D. Silva Dies at 92: Introduced News Helicopter. *Los Angeles Times*, [online]. Available at: http://articles.latimes.com/2012/dec/07/local/la-me-john-silva-20121207 [Accessed 13 September 2017].

Rule, T. A. (2015). Airspace in an Age of Drones. 95 B.U. L. Rev. 155. Available at SSRN: https://ssrn.com/abstract=2482567

Schroyer, M. (2014) Drone Lawyer: First Amendment Right to Photograph Extends to Drone Journalism. *Professional Society of Drone Journalism*, [online]. Available at: www.dronejournalism.org/news/2014/2/drone-lawyer-first-amendment-right-to-photograph-extends-to-drone-journalism [Accessed 9 June 2017].

Siddique, H. (2017) Drone Complaints Soar as Concerns Grow over Snooping. *The Guardian*, [online]. Available at: www.theguardian.com/technology/2017/apr/03/drone-complaints-soar-as-concerns-grow-over-snooping [Accessed 9 June 2017].

Sneed, A. (2015) 'So Your Neighbor Got a Drone for Christmas', Scientific American, www.scientificamerican.com/article/so-your-neighbor-got-a-drone-for-christmas/ [Accessed 22 December 2015].

Storyhunter.com. (2016) Storyhunter Guide to Commercial Drone Regulations around the World. [online]. Available at: https://blog.storyhunter.com/storyhunter-guide-to-commercial-drone-regulations-around-the-world-5795c31165d9 [Accessed 9 June 2017].

Voss, P. (2013) Rethinking the regulatory framework for Small Unmanned Aircraft: The case for protecting privacy and property rights in the lowermost reaches of the atmosphere, http://ieeexplore.ieee.org/document/6564687/?reload=true [Accessed July 2013].

surveillance "forces journalists to think and act like spies" (https://cpj.org/2015/04/attacks-on-the-press-surveillance-forces-journalists-to-think-act-like-spies.php). The algorithms of networked technologies coupled with big data processes permit the surveillance and aggregation of any kind of data to an extent that was previously unthinkable. This fact has led to an extended focus on the surveillance of journalists and the need for journalists to use encryption and other methods to protect their sources.

From a public perspective, though, one might argue that if surveillance is watching people and keeping track of how they behave, that is precisely what journalists do. More specifically, by focusing on keeping people in power, in particular, accountable for their actions, journalists engage in surveillance for the good of democracy. To be sure, systematically watching people to hold them accountable is not the only thing that reporters do, but watchdogging is an integral part of the processes of producing news. With new technologies, reporters have an array of new surveillance tools at their disposal, just like the police, the government and other institutions do.

As members of the Big Data society, in which the algorithmic coupling of public and private data retrieved from business as well as from government has become the new routine, journalists are of course part of the process. In the new game of surveillance, focus has moved from discipline to control (Lyon 2014). Surveillance is a necessary means used for border control and crime detection. The increasing numbers of terrorist attacks and cyber violations are valid arguments for legalizing even more high-tech surveillance and metadata collection.

In this larger picture of transparency and accountability, then, the nature of journalistic surveillance appears to be more of an ad hoc phenomenon related to investigative reporting than a feature of journalism in general. News journalists use a multitude of techniques for finding out what people are doing or have done, and the methods they as humans apply have become pretty advanced. At the same time, increasing amounts are invested to eliminate the human factor in journalism, too. The robotization and automation of journalism services are considered by many to be a necessary means for journalism and news media to survive the challenges of vanishing business models.

The negative connotation of surveillance seems to come not just from watching itself, but from how the watching is done, by whom and for what purposes. When it comes to news media, many forms of reporting that contain elements of surveillance are considered perfectly acceptable and wanted. Journalistic transparency, journalistic accountability and journalistic surveillance are all aspects of freedom of expression. These cyclic processes are tied together by journalism's codes of conduct and are as such in a constant contextual flux. What is considered acceptable behavior by, for instance, paparazzi in one country at one point in history might be condemned in another context and vice versa. Thus, when it comes to drone journalism, new questions are constantly raised about how drones are used, by whom and for what purpose. What is ethical, and what is not ethical drone surveillance in journalism? Transparency and openness could actually make the difference here. Paparazzi use of drones is clearly often interpreted as unethical, but as Tina Turner's wedding in Switzerland in 2015 demonstrated: when drones were not allowed, the paparazzi simply turned to the good old method of using helicopters. As such, the relationship between celebrities and paparazzi is a never-ending cat and mouse game.

Recognizing that news media and drone journalism involve surveillance situates them as a constituent of surveillance societies. Surveillance societies are societies in which information technologies embed the collection of records of human behavior into the infrastructures of daily life. David Lyon describes surveillance society as follows:

> To think in terms of surveillance society is to choose an angle of vision, a way of seeing our contemporary world. It is to throw into sharp relief not only the daily encounters, but also the massive surveillance systems that now underpin modern existence. It is not just that CCTV may capture our image several hundred times a day, that check-out clerks want to see our loyalty cards in the supermarket or that we need a coded access card to get into the office in the morning. It is that these systems represent a basic, complex infrastructure which assumes that gathering and processing personal data are vital to contemporary living. Surveillance is part of the way we run the world in the twenty-first century.
>
> (Lyon 2008)

Lyon goes on to explain that although surveillance society is often thought of as something sinister, it is "better thought of as the outcome of modern organizational trends in workplaces, businesses and the military." And yet it should be added that even in a globalized society, the effects of government surveillance of individuals vary along a broad spectrum. Depending on the level of press freedom in each country, the general impact of government surveillance on its citizens varies. The actual implications of surveillance data collection carried out in totalitarian regimes, such as North Korea, are somewhat different from the ubiquitous capturing of multiple layers of data in democracies such as those of the Nordic region. Even if available technologies for surveillance do not differ much from country to country, there seems to be a broad spectrum of ways that surveillance data are used, or also not used, but possibly only archived, in different countries.

In a journalistic context, though, the civic use of drones may be seen as another, perhaps predictable, step deeper into surveillance society. Unmanned aerial vehicles, available to anyone, are quickly becoming a natural part of many indoor and outdoor human activities. The use of drones is a new addition to what Lyon describes as the "complex infrastructure which assumes that gathering and processing personal data is vital to contemporary living." This trend will likely continue, with drones becoming a familiar presence in many other contexts.

With new gadgets there is always first a fear that along with more information and cases comes more understanding, and often acceptance. In time, drones will be as ubiquitous as smartphones. Who wants to prohibit the use of smartphones anymore in hospitals or bars? Maybe smartphones still have to keep quiet in schools for a few years, although they are increasingly integrated with new forms of teaching and learning. But drones are introduced and used in schools.

Although Lyon would have us think somewhat neutrally about a surveillance society – "as a modern organizational trend" – the threats to democracy and to democratic citizenship are palpable. Being watched all the time tends to inhibit basic liberties and freedom of expression (Richards 2013). Drones add to the infrastructure of watching. As individuals become aware of the use of drones, the presence of drones may be normalized, with awareness of being watched from the sky contributing to the sense that there is nowhere to hide. Whereas in the past, one might have felt a sense of privacy in going for a walk in the country, now such behavior might be recorded from a "camera in the sky" and consequently may have consequences at some time in the future.

Ironically, even though news media are part of a surveillance society, they are, at the same time, an essential component in counterbalancing the effects of surveillance. In surveillance societies, transparency is seen as an important antidote to the ill effects of surveillance. The bitter taste of being watched all the time is lessened when those who are watched know when, where, how and by whom they are being watched. One of the functions of news media and journalists is thus to watch the watchers. So news media and drone journalism are both surveillance *and* transparency. They are part of the watching (with its negative connotations) and part of the antidote (with its positive connotations). They watch and they watch watchers; they watch what people do and they watch what is done to people.

The preceding suggests that the use of drones is a new part of the preexisting role of news media and journalism. Does the use of drones add something unique? Is there something about the reach, the visual perspective or something else that requires special attention to the impact of drones on journalism?

Privacy

The danger of framing surveillance as simply "part of the way we run the world in the twenty-first century" is that it suggests there is no point in being concerned about privacy. Indeed, the current trend is to think that "privacy is dead." Nevertheless, privacy is still an important value even in surveillance societies. This can be seen in legislation (consider the EU right to be forgotten), in court cases and in some market behavior suggesting consumers prefer privacy. Privacy protection affects what news media and journalists do, that is, it constrains the behavior of news media. For example, law and professional norms limit what journalists can do in obtaining information, how they deal with sources and what they can reveal in stories.

As usually happens when new and disruptive technologies take off in a market, practice comes first and appropriate regulations later. Although aerial regulation to date has not been directed explicitly at news media (Holton et al. 2015), legislation is likely to affect journalistic practice (see Chapter 3) and could affect the nature of threats to privacy.

Perhaps the most daunting question is whether drones, on a global scale, should be deployed for any event that is of public interest, or whether there are places or types of events to which news drones should never be sent. Within the already existing aviation regulations, which specify the altitudes and areas that drones are allowed to operate in, mini-drones and drone swarms prompt new dilemmas. For example, what about covering private funerals if the drones are noiseless and invisible micro air vehicles (MAVs)? What about events involving children, or events held on private property? Questions of this kind are not new; in the past news outlets have had to establish policies with regard to new technologies and where reporters can and can't go. Drones pose a new version of the issue. In the new version, policies must take into account the somewhat unique ways in which drones can intrude on privacy. As the technologies are constantly evolving and equipped with more sensors, the complexity of these issues seems to be almost endless.

The uniqueness of drones, that is, the new feature that drones add to journalism, is range, sensoring and conditions of operation. Drones can gather images of events that neither humans nor other aerial vehicles such as helicopters can obtain. This has meant that news media can cover difficult-to-reach places, places where it would be dangerous to send reporters (see also Chapters 1, 3 and 5).

From a privacy perspective, it means that drones can go places that overcome "natural hurdles" that in the past allowed individuals to be left alone. In a sense, nature and architecture used to limit where journalists could go. For example, hurricanes prevent airplanes from immediate presence at the primary site of devastation. More simply, doors can be closed, window shades can be drawn and roads can be blocked, keeping reporters out of places where people don't want them. Drones overcome certain natural hurdles. They can go long distances from where they are launched and can fly above geographic hurdles. And as drone technology continues to develop, the range of operation is likely to extend further and further.

The increased range of where news media can go via drones poses a challenge to privacy in the obvious sense that individuals who want privacy have to do more; they have to take different kinds of steps to protect their privacy. On the other hand, most drones are easily detected, because of the noise of the rotors and their colorful lights. Down to a certain size and problematized by the emerging MAVs, one can argue that surveillance by drones is hard to make secret. Surveillance by military drones exemplified by the Global Hawk is at a different level. But new dilemmas arise now that the boundaries between the civic and the military are getting blurred. The smallest military camera drone until recently, the Black Hornet, weighs 17 grams (a little more than half an ounce) and is sold to more than 70 countries for civic and military purposes (www.tu.no/artikler/her-tar-forsvaret-i-bruk-sine-nye-black-hornet-droner/276145). This little gadget is capable of staying in the air for 25 minutes and flies between three and five meters a second, up to 1.6 kilometers above the ground station.

Although increased range is one of the most significant features that drones add to journalism, the threat to privacy derives not just from range but from the combination of increased range and advanced cameras and other sensors that record multiple layers of data.

Another significant feature of drones is that they produce new angles of vision. This means new ways of thinking about and understanding events and activities. New forms of understanding of human behavior have the potential to lead to new types of privacy intrusions. Importantly – though not unique to drones – the visual images that are produced by camera drones have permanence and reproducibility. The angles of vision produced by drones could mean new ways of thinking about human mobility and new ways to group individuals.

When it comes to drone journalism there is also an issue with regard to secondary information. When news media send out drones to cover an event, the drones will gather data, such as visual imagery, that were intentionally sought after. At the same time, they will gather other sensor data that were not sought after but picked up as part of the intentional data collection. The latter is referred to here as secondary data collection. As an example of secondary data, consider the hypothetical case of a drone deployed to gather pictures of a natural disaster; the footage collected includes individuals engaged in illicit behavior or simply behavior that the individuals would not want made public (even though legal).

Secondary information is not, of course, unique to camera drones. Secondary data may be gathered any time a camera is used. However, with camera drones, there is a difference in the kind of secondary data that may be picked up.

So drones have an increased range, and those who operate the drones are in the possession of the power, durability and reproducibility of visual imagery. In order to address concerns about privacy, then, news media will need policies on how they use drones, and they will have to be transparent about their usage. They will have to be transparent about when, where and how they are using drones.

The tough issues here will have to do with drawing a line between private and public, and when consent is needed and when not. In the contemporary media landscape, and particularly because of the digital surveillance capabilities of large tech companies and nations, these issues are much debated. Drones represent only one of very many digital communication tools. To be a bit provocative: seen from an aerial perspective, how could drone operators possibly know about the privacy level of an event? Seen from above, it may seem as if individuals have made no effort to make the event private. In that case, should news media presume an event is private unless told otherwise, or presume events are public unless told otherwise? And how can legislators possibly cope with questions like that?

Trust

The reliability of the information a media outlet or individual journalist produces, is of fundamental importance. Without the trust of their users, news media and journalists cannot survive. As already suggested, transparency is one of the key ways in which news media build trust with their audiences. Some of the challenge of drone journalism overlaps with the general challenge of digital imagery. That is, digital imagery is malleable and can be fabricated and/or modified. There is, however, an additional element with drone journalism insofar as there is no eyewitness to the filming. As Gynnild explains, "Since the early days of journalism, eyewitness accounts have been considered crucial in establishing authority and ensuring the credibility of news stories" (2014a, p. 334). When journalists and camera crews take pictures or film events, they are eyewitnesses to what is simultaneously recorded. These eyewitnesses are absent when drones record events.

News audiences may put their faith in video recordings. In fact, visual imagery may be more compelling and trust-eliciting, but the reality is that digital media are malleable. Although perhaps not intended, the presence of reporters and camera crews in non- and pre-drone journalism serves as a check on the reliability of digital images. In drone journalism, that checking is missing. Hence, news media have a bigger burden of establishing trust. An interesting question is whether checked authenticity of images and videos increase trust.

The challenge of establishing trust in drone journalism is compounded by the fact that drone journalism often means citizen journalism. In fact, if regulation increasingly restricts how the news media can use drones, that increases the likelihood that news outlets will use video footage provided by private citizens. Actually, the situation is quite paradoxical. Media professionals are sometimes not allowed to use drones, but hobbyists are. Amateurs are typically doing exactly the same job the professionals could do. Would society be just as ready for citizen doctors taking care of the patients, or citizen lawyers ruling in the courts? Furthermore, serious ethical questions arise if citizen drone journalists' outputs are used in crisis reporting, in hazardous environments, because professional, trained journalists are not allowed. Holton et al. (2015) describe an incident of this kind:

> When a gas leak caused a big explosion in New York in 2015, one of the first persons that came to the place was a business systems expert. He brought a drone and started filming above the blast zone after getting permission to film from relevant authorities. The 30 minutes of video provided early documentation of the leak, and the footage was sold to a number of news outlets. The details of the leak could not have been documented by helicopter.

Ordinary citizens who provide news imagery to news outlets from their smartphones are not new to news gathering. The novel feature with drones is that the footage is more extensive and more distant from the individuals who capture it than we are used to with video clips from, for instance, smartphones. Once again, there is no eyewitness to the event captured by the drone. This practice poses a challenge for news media in terms of guaranteeing the reliability of the imagery, and in terms of being transparent about their practices. From the perspective of privacy and surveillance, citizens using drones means citizen-to-citizen surveillance, which is different from journalistic surveillance.

Tremayne and Clark (2014) identified several significant cases in which private citizens contributed video footage from their drones (see also Chapter 3). This tendency may indicate a trend toward a kind of crowd-sourced news that might have enormous implications for the future of news media and its accountability. While journalism as a societal institution and with strictly practiced codes of conduct is supposed to guarantee the truthfulness and reliability of a story, hobbyists are not bound by the same practice. To the extent that the news media outsource the piloting of drones, their new locus of attention will be on fact-checking imagery shot by hobbyists and ordinary citizens.

Wikipedia might be the appropriate model for the future of news media. How will news media be able to certify the reliability of citizen contributed videos? News media will need to develop guidelines and principles for when they will and won't use video footage that is contributed by private citizens. Suppose the drone was sent into territory from which it was prohibited to go. Suppose the footage was taken secretly and without the consent of those involved. Suppose the footage contains secondary information that will be harmful to those who are innocent. Policies and citizen codes of conduct will be necessary independent of the law.

Autonomous drones

If civilian drones develop in the future along the same lines as military drones, they are likely to become more autonomous. This means that media users will be able to program drones to travel longer distances and to make decisions for themselves about where in particular to go and how to cover particular events. Instead of being remotely controlled or even pre-programmed to follow certain paths and record as they go, drones may be given more generic orders and allowed to make decisions depending on what they find as they operate. Programming such drones will not be an insignificant task. Based on what is happening in the development of military drones, one can imagine autonomous drones that have been programmed to identify the important locations within a terrorist attack or a natural disaster and to select the best height and angle of vision. One could also imagine that face recognition software is used to follow certain persons.

In the case of military drones, a number of ethical issues have arisen concerning the actions of the drones. However, even when the behavior of drones is aimed at hunting for enlightening news imagery, issues of responsibility are significant. No matter how intelligent, autonomous drones will have been programmed by humans to behave in certain ways, and humans will have decided to launch the drones. Hence, responsibility will always stay with the humans who decide how to program the drones and when to launch them.

References

Allen, D.S., 2008. The trouble with transparency: The challenge of doing journalism ethics in a surveillance society. *Journalism Studies*, *9*(3), pp. 323–340.

Craft, S. and Heim, K., 2009. Transparency in journalism: Meanings, merits, and risks. *The Handbook of Mass Media Ethics*, pp. 217–228.
Culver, K.B., 2014. From battlefield to newsroom: Ethical implications of drone technology in journalism. *Journal of Mass Media Ethics*, *29*(1), pp. 52–64.
Dennis, K., 2008. Keeping a close watch – the rise of self-surveillance and the threat of digital exposure. *The Sociological Review*, *56*(3), pp. 347–357.
Diakopoulos, N., 2014. *Algorithmic Accountability Reporting: On the Investigation of Black Boxes*. Tow Center for Digital Journalism. https://doi.org/10.7916/D8ZK5TW2.
Diakopoulos, N., 2015. Algorithmic accountability: Journalistic investigation of computational power structures. *Digital Journalism*, *3*(3), pp. 398–415.
Eide, M., 2014. Accounting for journalism. *Journalism Studies*, *15*(5), pp. 679–688. Published Online, 18 March 2014. http://dx.doi.org/10.1080/1461670X.2014.891856.
Gynnild, A., 2014a. The robot eye witness: Extending visual journalism through drone surveillance. *Digital Journalism*, *2*(3), pp. 334–343.
Gynnild, A., 2014b. Surveillance videos and visual transparency in journalism. *Journalism Studies*, *15*(4), pp. 449–463.
Holton, A.E., Lawson, S. and Love, C., 2015. Unmanned aerial vehicles: Opportunities, barriers, and the future of "drone journalism." *Journalism Practice*, *9*(5), pp. 634–650.
Jarvis, J., 2014. The ethical debate of drone journalism: Flying into the future of reporting. *Research Paper*, *475*.
Karlsson, M., 2010. Rituals of transparency: Evaluating online news outlets" uses of transparency rituals in the United States, United Kingdom and Sweden. *Journalism Studies*, *11*(4), pp. 535–545.
Lyon, D., 2007. *Surveillance Studies: An Overview*. Cambridge: Polity.
Lyon, D., 2008. *Queen's University, Canada, Talk for Festival del Diritto*. Piacenza, Italia, September 28. www.festivaldeldiritto.it/2008/pdf/interventi/david_lyon.pdf.
Lyon, D., 2014. Surveillance, Snowden, and Big Data: Capacities, consequences, critique. *Big Data & Society*, 1(2), Article first published online: July 9, 2014; Issue published: July 10, 2014.
Noorman, M. and Johnson, D.G., 2014. Negotiating autonomy and responsibility in military robots. *Ethics and Information Technology*, *16*(1), pp. 51–62.
Richards, N.M., 2013. The dangers of surveillance. *Harvard Law Review*, *126*(7), pp. 1934–1965.
Tremayne, M. and Clark, A., 2014. New perspectives from the sky: Unmanned aerial vehicles and journalism. *Digital Journalism*, *2*(2), pp. 232–246.
Wahl-Jorgensen, K. and Hunt, J., 2012. Journalism, accountability and the possibilities for structural critique: A case study of coverage of whistleblowing. *Journalism*, *13*(4), pp. 399–416.

5 Drones, teaching and the value of the explorative player-coach

Turo Uskali and Astrid Gynnild

Introduction

While the market for civilian drones – in all shapes and forms – has exploded in recent years, there are amazingly few courses offered on how to fly drones. More precisely, even if camera drones are now widely used by journalists on all continents, journalism programs seem to be lagging behind when it comes to integrating this disruptive tool into their teaching. By contrast, anecdotal evidence suggests that in disciplines such as geography, architecture, landscaping, agriculture and engineering an increasing amount of resources is invested into preparing students for the emerging drone future. Thus, we wanted to find out why and how a few journalism schools have truly flown out of the mainstream. We start with Finland and from there broaden the scope to early adopters in the United States and elsewhere.

The diffusion of civilian drones is particularly interesting to follow in a time when journalism education is under great pressure. As educators of students to an unknown future, we are obliged to adapt quickly to the new technological, economic and social-cultural changes that challenge the work roles and practices of journalists and other media workers. Simultaneously, teachers of journalism typically want to maintain all the core skills and virtues of the profession: being critical, writing excellently, using multiple sources and holding people in power accountable (Goodman & Steyn 2017; Hovden, Nygren, & Zilliacus-Tikkanen 2016; Terzis 2010).

Depending on assessment perspective, journalism educators are typically expected to fulfill different roles in their teaching: they might be judged as university professors concerned with critical thinking and media theory; they might be judged by their ability to focus on hands-on training of core journalism skills and values; they might be considered newsroom managers who should train operational competitiveness and speed. Or they are expected to be technological optimists who, in line with many editors and

DOI: 10.4324/9781315163659-5

CEOs, are convinced that new technologies are the utmost solution to most journalism challenges. Yet another option is to judge whether journalism teachers follow their entrepreneurial instincts and skills.

Mark Deuze (2006) summed up by suggesting that journalism education might choose between two major positions in society: the "follower" mode or the "innovator" mode. Based on the aforementioned literature, it appears that most journalism schools have traditionally chosen the follower role, emphasizing the core skills and sustaining old, conservative course modules; in other words, they play it safe.

Taking a role as a proactive facilitator of emerging trends and gadgets does not necessarily imply riskier risk taking, so to speak. It often tends to be overlooked that much risk taking associated with entrepreneurial approaches in real life do not apply directly to journalism education. An interesting difference is, for instance, that while entrepreneurs have to constantly keep looking for new ways of creating rapid and sustainable revenue, journalism teachers can experiment without risking loss of their jobs or loss of their colleagues or employees. The only thing teachers might risk is actually losing students if the journalism courses they teach do not appear up-to-date. Rather, this study indicates, in a time of constant change and tough production pressure, journalism programs at universities might be very well suited to creative experimentation and the nurturing of innovative mind-sets among students.

Journalism education has, just like other professionally oriented programs, been criticized for making students good at reproducing material but less trained in learning by doing. The Department of Communication at the University of Jyväskylä in Finland, however, has deliberately taken another stance and established a different kind of reputation. The journalism school in Jyväskylä was among the first in the world to introduce innovation to the curriculum: the first courses in innovation journalism were held in 2004 (Lassila-Merisalo & Uskali 2011). The allocation of resources to journalism innovation might partly be explained by teacher visits to Silicon Valley and Stanford University. Additionally, the department had a competitive advantage in being small and could easily and without risk change its bachelor and master programs in the desired direction.

With drones, the staff and students at Jyväskylä are again among the first in line. Students at this university were trained in using camera drones long before the drone issue was brought up in Finnish newsrooms. More than 30 graduate students, experienced in piloting drones, entered Finnish newsrooms after the first drone course was held in 2014. It is still a bit early to conclude in what ways their dronalism skills will affect Finnish media and society. But if we reason deductively, there should be a fair chance that the expertise in drone piloting will be considered an asset to the newsrooms.

The open-minded course design suggests that teachers and students engage in a peer-to-peer learning environment in which the speed of learning accelerates through rigorous experimentation.

Research question and challenges

Two burning questions to be investigated in this chapter are thus why drone courses are run in Jyväskylä and how they are facilitated. What was the philosophy behind this particular active learning approach, except that the university's Department of Communication clearly considers innovation journalism to be a part of its mission? The further discussion in this chapter is based on empirical data from the courses, such as curriculum, continuous dialoguing with students, student evaluations and reflexive discussions and notes carried out by the teachers.

A main challenge with the learning-by-doing approach of a new technology such as civilian drones is obviously the lack of knowledge and lack of access to relevant manuals on how to operate drones. Add to this challenge that by the time the first drone journalism course was held at the University of Jyväskylä there were no special rules for using unmanned aerial vehicles in Finland. Teachers and students operated indoors and outdoors and were free to experiment within amazingly wide frames. Even after the introduction of national regulations in October 2015 (Trafi 2015), the Finnish drone rules are still quite liberal compared to those of many other countries (Lauk et al. 2016).

Within such wide frames, what was then the best way to introduce new tools like camera drones to master students? That was the main question posed by the journalism teachers before the course started. After three years of experiential learning, the teachers are still asking the same question. Thus, a main aim of this chapter is to recapture, refocus and reflect on the intensive learning processes that actually took place during these courses. Admittedly, to begin with, the teachers did not take on any structured pedagogical approach except that, with an open mind, they started pondering the aforementioned questions. Following the theorizing of Peter Drucker on the entrepreneurial society (2014), it appears that the teachers, as well as the students, engaged enthusiastically in an intuitive learning-by-doing approach.

Experiential learning as a theoretical framework

In order to understand the learning processes that were constantly going on in parallel with an intensified focus on practicalities, we lean on the theory of experiential learning developed by Kolb (1984) and the theorizing of learning space (Kolb & Kolb 2005). As previously discussed by Gynnild

(2016), the experiential learning approach has widely influenced university teaching across disciplines but is in particular applicable to theory-practice approaches of Nordic journalism programs. It should be noted here that the term experiential is not a writing error but a concept developed from experimentation, on the one hand, and experience on the other.

Following Kolb's experiential learning theory, then, learning is cyclical in the sense that it follows a pattern that repeats itself. First, students learn from directly experiencing the events. Second, students learn from reflecting on these experiences. Third, students learn from conceptualizing these experiences, and fourth, students test what they have learned by applying the new knowledge. At every step of the cycle, new learning ideally leads to constantly progressing reflections and discussions that feed the further perception and understanding of what is going on.

Learning by experiencing

In the following paragraphs, we provide some descriptive examples of concrete challenges that had to be resolved by teachers as well as by students in Jyväskylä when experimenting with drones. What lessons were learned, and how were these issues handled by the educators? The challenges concern, in the first instance, practical issues such as types of drones, weather limitations, flying space and flying practice needed. They add to the understanding of what is needed for journalism experimentation to take place. Next, we compare these experiences with parallel processes going on in higher education in the United States. By analyzing these data in turn, we arrive at three aspects that journalism teachers have in their roles as leaders of active learning processes of future journalists.

With hindsight, the initial use of drones in the Jyväskylä journalism program did bring with it a good deal of risk taking. The course started out with Parrot AR, a French drone that in 2010 had become the first civilian bestseller (Goldberg, Corcoran, & Picard 2013). A main difficulty with this lightweight drone model was that even mild winds easily caused disruptions. One of the first student flights ended on the rooftop of a block of flats in the city center of Jyväskylä. After several minor crashes by other students, the drone was wrecked, and teachers started looking for more advanced drone models. The first model of Parrot AR was still a toy, not a proper tool for journalism, as experienced by other drone educators as well.

In 2015, the master program bought its first DJI Phantom 2, which was assessed in this way by *The Economist* (2015): "That brought professional-quality aerial photography within the reach of general users." Later the same year the teachers invested in two Chinese Yuneec's Typhoon Q500s, followed by another new model DJI Phantom 3 4K. In 2017 two DJI Mavic

Pros were added to the fleet. A lesson learned is that to improve drone journalism education, one has to be willing to invest constantly in new drone models. A broad rule of thumb is that one drone will be lost per course.

Another issue that had to be resolved on the spot was the question of time needed for flying. For example, in order to be a helicopter pilot, a minimum 40 hours is required in many countries, but actually most candidates practice more (Perritt & Sprague 2017). Some experts have argued that formal training requirements for using drones may not be necessary at all. Perritt and Sprague (2017) emphasized that one has never needed a license to fly model airplanes or to use chainsaws or lawn movers. More importantly, according to Federal Aviation Administration rules (FAA 2016), and upcoming European Union rules on unmanned aerial vehicles, at least a theoretical online test is needed for piloting drones for journalistic purposes.

The minimum requirements for the University of Jyväskylä students, before starting operating the drones themselves, were to participate in at least 80 percent of the classes (10 hours of theoretical lecturing), and have at least 60 minutes of practical flying time under the supervision of the instructors. All flying practice sessions were held at the university's sports field.

After the first direct experiences, and much discussion and reflection among the teachers, it became evident that drone flight practice did require clear prescriptions on every step of the process. This learning lesson is supported by the findings of Perritt and Sprague (2017). Clear structures are needed for a) preflight preparation and procedures, b) takeoffs and landings, c) hovering maneuvers (in slow/fast modes), c) short-term missions, d) aerial image and video practices, e) emergency operations (including using the autopilot) and f) post-flight procedures, which include, for instance, documentation in the flight log and recharging the batteries.

The next learning lesson concerns weather conditions, which matter greatly in drone piloting. In general, the best time for outdoor practicing, especially in the Nordic countries, is from the late spring to summer and early autumn. In the case of days that are too windy or rainy, the practice sessions were sometimes moved indoors, to a TV studio, and the outdoor drones were switched to mini-drones. Failing safely became important when using the toy-like mini-drones. These drones were particularly difficult to operate; a one-minute flight was record-breaking. Also using the drone piloting simulators offered by the manufacturers was helpful when outdoor practice was not possible.

After four weeks of lectures and practice, students were able to test what they had learned. The main homework was to produce a drone journalism story, in pairs, without any help from the instructors. The themes for the short video stories varied from the city dump and a Pets' Day cavalcade to students' festivities and sport events. All 25 students went through the lectures and practical training sessions, but four students did not return the drone course work (a short drone video and a course feedback report) in order to pass the course and get a grade. In general, it is typical in advanced, MA-level voluntary journalism courses that some students will drop out during the course because of coursework overload from other courses.

As the educators engaged actively in the learning-by-doing-approach themselves, a next step was to collect written feedback systematically from the students. The educators were hungry to get constructive critiques in order to develop the course further. All the students who answered the survey (N = 21) were overwhelmingly positive and considered the drone journalism course very successful. One student commented that the course served as an "eye-opener." Another student said, "The best course ever: educative and fun." One student even postponed her graduation because she wanted to take the course. All the students also wanted to recommend the course to other students and advised instructors to continue the course, despite some setbacks.

On the more critical side, most students wanted to have more time for piloting the drones. Sixty minutes was not perceived as sufficient to ensure basic skills for operating the drones safely. The students also suggested more challenging practicing places than the open campus sports field; some students even suggested a special test arena for drone practicing. Finally, according to the survey, too many drone videos were watched during the lectures.

Indeed, calculating the numbers of broken or completely wrecked drones, the dropout rate, and the students' feedback, one can argue that at a practical level, the basic course model still needed some improvement. Another rather obvious rule of thumb is that the more practice one gets, the better one becomes at drone piloting; one hour of flight time was obviously not enough for preventing crashes. Therefore, in 2017 the new minimum drone piloting time was extended to 60 minutes in the sports field and 60 minutes in more challenging environments – all in all, two hours. Moreover, in the current drone courses at the University of Jyväskylä, the students have to pass two tests in order to get a "license" to operate the drones by themselves: one online test and one practical flying test. Passing the drone journalism course will give the students the permit to use the drones for other journalism courses as well. In this way, the educators hope camera drones will be just a new tool to students' tool pack for visual storytelling.

The frames for drone experimentation at the University of Jyväskylä have changed to reflect new Finnish drone rules. The regulations require that the department inform the Finnish aviation authorities about all drone operations. A drone operator's manual was written, and educators have to log all drone flights. Special drone insurance is purchased. Every time teachers and students want to use the drones, they need to ask the City Airport for permits by phone. The City Airport also wants to be informed when operations are finished. A special app published by the Finnish aviation authorities in 2017 includes updated information about the no-fly zones and other restrictions. The app uses the location of the user and displays the current situation in colors: red means no-fly-zone or call the nearest airport to check whether it is possible to operate the drone in the area. Green allows all kinds of drone activities.

Altogether, the new governance of unmanned aerial vehicles in Finland means that drone journalism education constantly adapts to new outer frames for experimentation. In a sense, one might conclude that stricter regulations point in direction of a more responsible model of drone journalism in higher education. But do these regulations, together with continuous reflections on one's own experiences, necessarily ensure that drone journalism education becomes responsible? Before we continue this discussion, we want to discuss the pioneering drone activities in American journalism.

Lessons learned in the United States

The educative lessons learned in the US are based on experiential learning as well. Whereas educators in Finland engaged in developing a Finnish model for dronalism, the Drone Journalism Lab at the University of Nebraska-Lincoln (Los Angeles Times 2011) has become a leading institution for facilitating drone journalism in the US. Its founder, Professor Matt Waite, was among the first to realize the importance of camera drones for journalism. Another early mover was Chris Anderson, *Wired* Magazine's editor-in-chief. He started to customize drones with his children in 2007 and later moved from journalism to become a drone entrepreneur and do-it-yourself community facilitator (Anderson 2012).

As previously innovative journalists and devoted drone enthusiasts, Waite and Anderson exemplify what Peter Drucker would call leaders as explorers. In journalism, they also represent the decreasing but very important group of creative news professionals, constantly hunting for news in the sense of being first with the latest – in new fields. The Drone Journalism Lab, for instance, obviously took an exploratory stance from the beginning. A main aim of the Lab was to search for answers to new questions by doing and experimenting. For example, Waite's first flying attempts with the Parrot AR ended in many indoor and outdoor crashes. The experiments were thoroughly documented in Waite's detailed reports on the incidents:

> At about 5:30 p.m. CST Dec. 28, 2011 during a test flight to gauge the effectiveness of a new camera mount, the drone operator lost control

> of the aircraft while it was flying over his house. The drone was flying away from the operator and not responding to commands to turn away from the house. The operator panicked and the emergency shutdown signal was triggered causing the drone to crash onto the roof, slide off the back of the house and fall two stories to the ground below.
>
> (Waite 2011b)

After his analysis of the main reasons for the crash, Waite reflected and instructed that one should not fly at night, avoid panicking and find control systems other than smartphone and Wi-Fi. Later, Waite added to the list that one should have a budget for the replacement parts. The crash damaged the front-facing camera and the landing bed. After the first weeks of testing, Waite concluded, "Crashing is a part of this, especially starting out" (Waite 2011a).

In 2013, the second drone journalism program in the US was started at the Missouri School of Journalism in collaboration with the University of Missouri Information Technology program. The project's goal was to discover how best to utilize drone technology in the field of journalism.

A striking feature of the American education initiatives is the intensive production of blog posts. Similar to that of the Drone Journalism Lab, the website www.missouridronejournalism.com has become one of the main news curators on drone journalism. Blogs typically serve as valuable outlets for reflection in action and collective sharing of experiences. The Missouri school's first drone journalism story was a clip about the unusual amounts of snow geese in the area. From the beginning, the blog posts of the Missouri project attracted international reporters from countries such as South Korea to observe the program on the spot (Garcia 2013).

Contrary to the first Nordic experiments with drones in journalism, which were hardly mentioned in blogs, the transparency of the American drone educators soon got them into trouble. The blog reports led to much attention from other bloggers and from the news media. Rumors and false information were circulated especially via conservative blogs – for example, that the Environmental Protection Agency (EPA) was using drones to spy on feedlots in Nebraska. According to Pham (2013a), the conservative lawmakers teamed up with the American Civil Liberties Union in legislatures all over the country. As a result, by the summer of 2013, six states successfully passed anti-drone laws, and 28 states (including Missouri) had anti-drone laws active in statehouses. The two research and education programs on drone journalism in the United States, Nebraska and Missouri, were subsequently halted by the authorities because they lacked special permits (Pham 2013a, b; Waite 2013). Drone journalism education in the US was halted from 2013 to 2016. In this period it was unclear on what premises, if any, teachers and students were allowed to operate drones.

After more than two years in "stealth mode," in October 2015, The Missouri drone journalism program again started to post its operations publicly on its blog. During the silent time, the program adjusted their physical learning spaces; they felt forced to continue flying practice indoors in a livestock arena. Moreover, the educators and researchers found a new creative outlet by going on reporting trips abroad, for instance, two weeks of field reporting in Costa Rica, and later in Zambia. (Missouri Drone Journalism 2015; Shaw 2015a, 2015b).

"Schools and universities are incubators for tomorrow's great ideas, and we think this is going to be a significant shot in the arm for innovation," said an FAA administrator in a BuzzFeed news story in May 2016. According to the FAA, the students were then allowed to use the drones for their schoolwork in high schools and colleges. The federal regulators defined drone schoolwork as a hobby or recreation, not as a commercial activity. Because teachers are paid, their use of drones is treated differently. In the story, Matt Waite argued that "the prohibition on teachers flying drones puts serious limits on the new policy's efficacy" (Shaban 2016).

Similar to many other countries, drone programs are spreading across universities and colleges in the United States. In late 2017, at least 16 different programs are up and running, but only two focusing on drone journalism (Dronethusiast.com 2017). The courses in higher education are typically supplemented by workshops, boot camps, hackathons and other activities outside of the universities. For instance, in China, People's Daily Website, Xinhua News Agencies and the China Daily website have joined forces with local drone makers to improve digital news coverage (Zheng 2016). In addition to face-to-face and hands-on educational modules, many online tutorials and simulator programs for a variety of drone models are offered by the drone manufacturers for self-learning (Perritt & Sprague 2017).

Discussion

From the many blog posts, videos and online discussions behind this study, there are, however, many voids when it comes to student experiences with drone journalism. The voices that dominate online educative forums are those of teachers – meaning teacher-managers, teacher-entrepreneurs or teacher-leaders, depending on the role that fits best. These front-runners apparently display very valuable qualities as leaders and spokesmen of journalism innovations, such as adopting drones as a newsgathering tool. Teachers in higher education are, contrary to managers and CEOs in mainstream newsrooms, not concerned about operational efficiency and competitiveness. Rather, they appear to be immersed in a kind of creative competitiveness in which the reward is not money but being first with the latest innovation.

As Drucker pointed out, building creatively competitive organizations requires curiosity first and foremost; asking questions is valued higher than providing answers.

Along such parameters, we argue that journalism programs in higher education have many assets, such as labs for journalism innovation. One of the most important assets is journalism teachers who passionately engage in innovation processes and intuitively lead students toward creative competitiveness. Following Drucker, leaders or teachers of creatively competitive organizations, exemplified also by some journalism schools, take on three different roles in the cycles of innovation (Brown 2016): The first role is that of the explorer, the person who is in front. The explorer leads by asking strategically purposeful questions that typically might bring great value to the institution. The second role is that of the gardener, which is more of a leading-from-behind; the gardener foresees and provides tools and spaces for purposeful experimentation and innovation. The third role is that of the player-coach, which implies to lead from the side without taking over what the students are doing. The player-coach is supposed to anticipate what problems may arise and follow and support the students as they actively engage in resolving the issues with which they are working. The liberty of journalism teachers lies in not having to worry about the revenue on a daily basis, unlike entrepreneurs in the industry.

From the data in this study, the outstanding position as journalism innovation schools in Finland as well as in the American states of Nebraska and Missouri did not evolve by accident. The institutions took on these roles because they nurtured teaching leadership directed toward creative competitiveness instead of operational efficiency. The opening up for journalism innovation first attracted more devoted teachers. In the next round, the facilitation of innovation opportunities represented by drones most likely attracts students who see drone expertise as a competitive advantage. According to Kolb and Kolb's theorizing on learning spaces (2005), it is crucial that members of a learning community are known and respected by faculty and by colleagues. They need to feel that they are allotted a space.

And yet we argue that when it comes to facilitate critical and questioning approaches to drones in a societal perspective, journalism schools still have much to learn. That is the case at least if we place the drone journalism courses in a responsible research and innovation (RRI) perspective. At this point in drone journalism history, it appears that educators are still mostly concerned with the practicalities of piloting drones as a newsgathering tool. There is little discussion about drones as a news beat, even though millions of Euros and dollars are invested in the skyrocketing civilian drone industry. Moreover, even though several thousand civilians in the Middle East have been killed by drone warfare, these actions go on with little interference from the news media. So the following questions arise: In what ways might students best be provoked to begin debating the many unresolved issues of military and civilian uses of drones? In what ways might responsible droning be taught in ways that take into consideration the potential bad uses of the new technology?

Based on the study of early drone adopters in education, we do not have a single good answer to these questions. But we would like to throw in a last idea from the side: now that the first phase of drone teaching in journalism is settled, would there be more space for critical thinking and critical dialogue on the opportunities and dilemmas of drones? This would require consideration of drones not just as a newsgathering tool for journalism explorers but as an increasingly important military and civilian industry that deserves to be covered more broadly as a news beat.

Whereas critical thinking is considered a basic communication skill to be learned in bachelor programs, Morris (2017) suggests that teachers in higher education should focus more on critical dialogue. She claims that critical dialogue is "an active group process and opportunity for students and faculty members to learn how to engage in civil, respectful, difficult conversations." She points out that when engaging in critical dialoguing, teachers and students will "tap not only into the cognitive domain, but also into our attitudinal and behavioral predilections. Critical dialogue is learned in community and serves the community, and the process can unite students and faculty members from divergent backgrounds and viewpoints around difficult, yet shared, issues and problems" (Morris 2017, p. 1).

As suggested in a previous study of journalism innovation (Gynnild 2016), in order for innovation journalism to expand as a news beat, journalists need to explore what innovation entails in practice. With the experiential learning about drones conducted in higher education, the journalists of tomorrow are getting a flying start technically. The question is this: In what ways might the competencies already achieved extend to integrate the difficult issues as well? Ideally, critical dialogue should be a perfect match for journalism students who are trained in posing direct and provocative questions of all types. The only dilemma – and one that is apparently increasing – is that journalism educators and their students might be immersed in their own drone experiments to the extent that critical thinking and critical dialoging on behalf of society are lost. Therein lies a challenge for journalism teachers as explorers, gardeners and player-coaches.

References

Anderson, C. (2012) How I Accidentally Kickstarted the Domestic Drone Boom. *Wired*, [online], 22 June 2012. Available from: www.wired.com/2012/06/ff_drones/ [Accessed 10 March 2017].

Brown, T. (2016) Leaders Can Turn Creativity into a Competitive Advantage. *Harvard Business Review*, 2 November.

Deuze, M. (2006) Global Journalism Education: A Conceptual Approach. *Journalism Studies*, 7 (1), pp. 19–34.

Dronethusiast.com (2017) *16 Top Drone Programs at Universities and Colleges*, [online]. Available from: www.dronethusiast.com/top-universities-unmanned-aerial-system-programs/#Accredited [Accessed 11 April 2017].

Drucker, P. (2014). Innovation and entrepreneurship. London: Routledge.

FAA (2016) *Operation and Certification of Small Unmanned Aircraft Systems*, [pdf]. Available from: www.faa.gov/uas/media/RIN_2120-AJ60_Clean_Signed.pdf [Accessed 2 April 2017].

Garcia, Z. (2013) Korean Journalists Observe a Class, Drones in Flight. *Missouri Drone Journalism*, [online]. Available from: www.missouridronejournalism.com/2013/04/korean-journalists-observe-a-class-drones-in-flight/ [Accessed 10 April 2017].

Goldberg, D., Corcoran, M. and Picard, R. G., eds. (2013) *Remotely Piloted Aircraft Systems and Journalism: Opportunities and Challenges of Drones in News Gathering*. Oxford: Reuters Institute for the Study of Journalism.

Goodman, R. S. and Steyn, E., eds. (2017) *Global Journalism Education: In the 21st Century: Challenges & Innovations*. Austin, TX: Knight Center for Journalism in the Americas.

Gynnild, A. (2016) Developing Journalism Skills Through Informal Feedback Training. In Hovden, J. F., Nygren, G. and Zilliacus-Tikkanen, H. eds. *Becoming a Journalist: Journalism Education in the Nordic Countries*. pp. 321–332. Gothenburg: Nordicom.

Hennigan, W. J. (2011) Idea of Civilians Using Drone Aircraft May Soon Fly with FAA. *Los Angeles Times*, [online]. Available from: http://articles.latimes.com/2011/nov/27/business/la-fi-drones-for-profit-20111127 [Accessed 10 March 2017].

Hovden, J. F., Nygren, G. and Zilliacus-Tikkanen, H., eds. (2016) *Becoming a Journalist: Journalism Education in the Nordic Countries*. Gothenburg: Nordicom.

Kolb, A. Y. and Kolb, D. A. (2005) Learning Styles and Learning Spaces: Enhancing Experiential Learning in Higher Education. *Academy of Management Learning & Education*, 4 (2), pp. 193–212.

Kolb, D. A. (1984) *Experiential Learning*. Englewood Cliffs, NJ: Prentice-Hall.

Lassila-Merisalo, M. and Uskali, T. (2011) How to Educate Innovation Journalists? Experiences of Innovation Journalism Education in Finland 2004–2010. *Journalism & Mass Communication Educator*, 66 (1), pp. 25–38.

Lauk, E., Uskali, T., Kuutti, H. and Snellman, P. (2016) *Droonijournalismi: Kaukoohjattavien kamerakoptereiden toimituskäyttö. ["Drone Journalism: Utilizing Remotely Piloted Aircrafts (RPA) in Journalistic Purposes"]*. Jyväskylä: University of Jyväskylä.

Missouri Drone Journalism (2015) *Class of Drones*, [Online Video], 15 September 2015. Available from: www.youtube.com/watch?v=D_8UipaLJGc [Accessed 10 April 2017].

Morris, L. V. (2017) Moving Beyond Critical Thinking to Critical Dialogue. *Innovative Higher Education*. doi: https://doi.org/10.1007/s10755-017-9413-z.

Perritt, H. H., Jr. and Sprague, E. O. (2017) *Domesticating Drones: The Technology, Law, and Economics of Unmanned Aircraft*. London: Routledge.

Pham, S. (2013a) Missouri Drone Journalism Program to Reconfigure Goals after FAA Letter. *Missouri Drone Journalism*, [online], 21 August. Available from: www.missouridronejournalism.com/2013/08/missouri-drone-journalism-program-to-reconfigure-goals-after-faa-letter/ [Accessed 10 April 2017].
Pham, S. (2013b) When Journalism Becomes a Game of Drones. *Mashable*, [online], 28 July. Available from: http://mashable.com/2013/07/28/game-of-drones-journalism/#m2uoU7ni.sqy [Accessed 21 March 2017].
Shaban, H. (2016) Students Can Now Fly Drones at School, FAA Says. *BuzzFeed*, [online], 4 May. Available from: www.buzzfeed.com/hamzashaban/students-can-now-fly-drones-at-school-faa-says?utm_term=.viVv2BeoZ#.pwdQ8wgDv [Accessed 13 April 2017].
Shaw, R. (2015a) Costa Rica Trip. *Missouri Drone Journalism*, [online], 16 November. Available from: www.missouridronejournalism.com/2015/11/costarica/#more-518 [Accessed 10 April 2017].
Shaw, R. (2015b) Drone Journalism Website Returns to Action. *Missouri Drone Journalism*, [online], 1 October 2015. Available from: www.missouridronejournalism.com/2015/10/drone-journalism-website-returns-to-action/ [Accessed 10 April.2015].
Terzis, G., ed. (2010) *European Journalism Education*. Bristol: Intellect, Limited.
Trafi Finnish Transport Safety Agent (2015) *The New Aviation Regulation on the Use of Unmanned Aircraft and Model Aircraft*, [online], 10 September. Available from: www.trafi.fi/en/about_trafi/news/3588/new_aviation_regulation_on_the_use_of_unmanned_aircraft_and_model_aircraft [Accessed 10 April 2017].
Waite, M. (2011a) Crash Report No. 1. *Drone Journalism Lab*, [online], 30 December. Available from: www.dronejournalismlab.org/page/19 [Accessed 17 March 2017].
Waite, M. (2011b) Lesson: Budget for Replacement Parts. *Drone Journalism Lab*, [online], date. Available from: www.dronejournalismlab.org/page/19 [Accessed 18 March 2017].
Waite, M. (2013) Drone Journalism, the Rules and the Way Forward. *Drone Journalism Lab*, [online], 21 August. Available from: www.dronejournalismlab.org/page/10 [Accessed 21 March 2017].
Zheng, L. (2016) China Daily Partners with EHang on Drone Journalism. *China Daily*, [online], 7 December. Available from: www.chinadaily.com.cn/business/tech/2016-12/07/content_27602940.htm [Accessed 16 April 2017].

6 Taking risks with drones

Responsible innovation pedagogy for media education

Lars Nyre, Frode Guribye and Astrid Gynnild

Introduction

The increasing demand for drone imagery in the news media requires that more journalists learn how to operate camera drones. In order for journalism to uphold its professional standards of accountability in visual news coverage, camera drone piloting is not a competency that should be outsourced to external companies. Higher education institutions should have a special responsibility for educating drone pilots with a practical grasp of the possibilities and limitations. The quest for qualified technological skills as well as competent judgment when it comes to news values and journalistic codes of conduct is increasingly important. At this point in media history, it has become evident that journalism and new technology are closely related and that the journalism profession profits creatively from the disruptive tools that are introduced. At the same time, students of technology, like our bachelor students in "New media," have much to learn from journalistic approaches and considerations of societal issues of importance.

The design experiment we report in this study explores two interconnected topics that can be formulated as two claims: First, when students are exposed to unexpected forms of risk in a new technology, they open up to creativity and subsequent reflexive exploration of the technology in question. Second, the experienced risk stimulates most students to display greater carefulness, accountability and responsibility when using the given technology. In order to find out more about these topics or claims, an existing smartphone programming course was reoriented to be applicable for semiautomatic drone flying. A 3DR Solo drone with its software was introduced as the semester's main tool for the students, their teachers and administrators. As teachers, we were excited to investigate to what degree students would experiment with, and get a sense of, ethical dilemmas of visual intrusion as well as the value of shots and sequences from a drone perspective. Our study is informed by the principles of responsible research

DOI: 10.4324/9781315163659-6

and innovation (RRI). In this approach, the research and innovation process should have desirable and useful outcomes for society as well as be ethically acceptable and environmentally sustainable (von Schomberg, 2012; Strand, 2015).

The chapter continues with a theoretical discussion of the relationship between technology, risk and learning, and focuses on the possible benefits of taking risks with drones in journalism and media education. A theory of responsible innovation pedagogy is presented; thereafter, the experimental method and its qualitative characteristics are outlined. The analysis is split into three sections dealing with the "risk experiences" of students, teachers and administrators, respectively. In the conclusion, we present four RRI insights relevant for students, teachers and administrators involved with drone flying in media education programs at higher education institutions.

Technology and risk

Risk is a central topic in modern theories on technology and society. The material characteristics of technologies, their weight, speed and force, often imply a risk of physical damage to its users as well as to things and to other people. A worker operating a power loom in an early nineteenth-century steam engine factory was at risk of having his fingers torn off by belts; the driver of a car without seat belts today risks his or her life. In our context, the drone carries risks due to its advanced functionality. When flown, it combines high maneuverability, camera vision, stability, long battery life and digital storage and transmission of signals. For mediation purposes, a host of different sensors can be attached to a drone, for example, photo, video, infrared camera and directional microphones; such sensors augment information capabilities in the desired direction.

How should risk be defined? In a pragmatic approach, risk is understood as something that is relatively predictable, that can be specified and in most cases avoided by taking safety measures. Antonsen (2009, p. 6) writes: "A risk analysis basically consists of what may go wrong, how likely it is that something in fact will go wrong, and the consequences involved if these things go wrong." Risk is closely related to the concept of "safety." Wold (2016) points out that the concept of safety is not to avoid or prevent something going wrong, but "to ensure that everything – or as much as possible-goes right" (Hollnagel, 2014, p. 23). Wold writes that this "includes a focus on everyday activities, not just accidents and mishaps, as safety is understood as the ability to succeed under expected and unexpected conditions alike. Newer perspectives are increasingly involving social, cultural and technological factors in a dynamic interaction leading up to unwanted events" (Wold, 2016, p. 26).

We can now specify the risks involved in using a drone as part of a university course in media education. Clearly, there is risk of material damage or injury to persons due to the flying movements of the drone; such risk is enhanced by the students' lack of experience with controlling the drone. This risk relates to being underneath and near to a flying drone, and requires safety measures as well as insurance. It is the responsibility of the pilot and the drone teacher to ensure that all rules are followed. In addition to the physical risk, there is an economic risk of liability for persons and institutions if something should happen due to lack of rule-following. Furthermore, media personnel are at risk of breaching privacy regulations when filming and/or publishing video clips. It can be ethically problematic to publish certain types of drone clips.

While there is risk, there is also gain. A drone pilot produces an extremely mobile video perspective and can add significant value to any visually oriented media product. It is clear from the start that teachers as well as students have much to gain from engaging with drones.

Learning and risk

Our approach to understanding learning and risk is anchored in a sociocultural perspective (Säljö, 2009). Learning is seen as process of mastery of conceptual and material artifacts. We approach these practices as relational to an infrastructure for learning. When a new artifact – in our case a drone – is introduced into the infrastructure for learning, it serves as a mediating artifact and the object of an emerging learning practice. Further, we see the infrastructure for learning as a combination of pedagogical, technological and institutional arrangements. More specifically, the students are supposed to control the risks and explore the creative potential of the drone in an emerging practice that should be nurtured further by teachers and supported by administrators. In the analysis, we observe the relationship between learning, pedagogy and rule-following, and try to describe their influence on the emerging learning practice.

Weilenmann, Säljö, and Engström (2013, p. 749) argue that "as the notion of literacy shifts towards participation and the ability to produce media content, rather than just consuming it, and as the tools for production become more powerful and diverse, the skills needed to participate will be increasingly medium specific." In our design experiment, the main aim is to expose the students to the challenges of using a drone as a means for shooting video that is suitable for publication in a serious media outlet, and to explore further ways of programming the drone for journalistic purposes. Risks of damage, injury and liability are natural ingredients in such a learning scenario. The design of the learning scenario is inspired by innovation pedagogy (Kettunen, 2011; Darsø, 2011). According to Kettunen (2011), innovation pedagogy has its theoretical roots in the pragmatism of John Dewey. This pragmatic perspective is aligned well with a sociocultural perspective on learning (Säljö, 2010). A common focus is on the experiential nature of learning and how it is mediated by the tools of the trade. Kettunen (2011) further argues that this particular approach is well suited and flexible enough to accommodate challenges that arise in education in the applied sciences. This argument resonates well with our approach, as we focus on the application of technology in a journalistic context.

Innovation pedagogy, as emphasized by Darsø (2011), aims to teach students to become more innovative or creative by giving them responsibility for a development process. This tradition supports a definition of creativity as the ability to produce work that is novel – original and unexpected – and appropriate, useful and adapted to task constraints (Sternberg & Lubart 1999). The ideal pedagogical approach is to find a balance between making students accountable and guiding them with creative support and deliverables. Our approach is also inspired by problem-based learning (PBL), "a student-centered pedagogical approach in which students learn by the process of solving an open-ended problem within a team. . . . The PBL system differs from traditional instruction in that PBL engages the student in constructing knowledge and the teacher role moves from a knowledge provider towards a learning facilitator" (Hmelo-Silver, 2004).

Responsible innovation pedagogy

As previously mentioned, this study is informed by a specific value-orientation, namely the responsible research and innovation (RRI) framework – an approach anchored to European policy processes and values. According to von Schomberg (2012), responsible research and innovation is:

> A transparent, interactive process by which societal actors and innovators become mutually responsive to each other with a view to the (ethical) acceptability, sustainability and societal desirability of the innovation process and its marketable products (in order to allow a proper embedding of scientific and technological advances in our society).

The RRI approach has been developed further by Stilgoe, Owen, Macnaghten (2013), who state that responsible innovation "means taking care of the future through collective stewardship of science and innovation in the present." They call for improved skills of anticipation. Researchers and organizations must ask *what-if questions*, and they must be able to adjust their course of action when they are faced with new knowledge. The EU Commission supports a number of initiatives in the RRI sector. Notably, there is an online toolkit for researchers who want to adopt the RRI ideals in their projects (RRI Tools, 2017).

In this study we combine the ideas of innovation pedagogy with those of responsible research and innovation. Such a combination can be labeled "responsible innovation pedagogy" and can be associated with recent explorations of RRI that are tailored to higher education institutions. A Horizon2020 project exists that specifically addresses higher education institutions and RRI (see HEIRRI, 2017). We aim to contribute to the discussion; in the conclusion of this chapter, we summarize the insights from this experiment in four principles of responsible innovation pedagogy.

In order for a pedagogical arrangement to qualify as RRI, students, teachers and administrators must be engaged in an explorative, collaborative learning process that is not top-down. If the pedagogy works well, students are supposed to learn how to make a certain product and create their own knowledge along the way. The creative work is conducted in groups with great freedom of exploration but with strict rules for iterative development and time-boxing in order to deliver a quality product on time (as an exam deadline). Students are likely to learn how to collaborate in professional teams. Note that the direction of the course is adjusted if unexpected problems occur, or, in our case, if the perceived risk becomes too high.

The RRI approach has universal merit. Teachers are always concerned with finding ways to make students engage in creative learning activities in which they (students) design and create solutions, while simultaneously encouraging critical reflection on the implications and potential of the given design. We chose collaborative group work and open-ended problem solving to engage students in creative activities and critical reflection. Such course designs require a focused fostering of social bonds. In order to collaborate well, students need to experience a mutual sense of equality and fairness in the groups. Without such trust and acceptance of different specialties during the creative work, there would be little chance of the result becoming responsible.

Design experiment method

In the learning sciences, a long-standing tradition exists for conducting design experiments (Brown, 1992) and design-based research (Barab & Squire, 2004; Collins, Joseph, & Bielaczyc, 2004). On a general level, such studies involve making interventions in existing educational settings by introducing new technologies in concert with a deliberate pedagogical approach to induce change in learning practices while systematically studying the implications of such an intervention. In other words, changes are done to mediating technologies and the organization of learning activities inspired by a given pedagogical approach.

We basically inserted a drone into an established course module and then dealt with and documented the trajectory of events. The course took place in the spring semester of 2016 at a university in Norway and covered 10 ECT. The module has existed for years, and its purpose has been to explore a new technology for journalistic purposes each year. Last year, the topic was sound media for Samsung smartwatches; in 2014, it was sound media for Android mobile apps. Neither the university teachers nor the students had any experience with drone flying before the courses started. We hired an external drone teacher, who runs a professional company for drone filming and teaching. A 3DSolo drone and four small drones, mostly for indoor training, were used in the course. The students got four half-day sessions in which they explored flying the drones indoors and outdoors.

In this design experiment, we worked in accordance with three principles of innovation pedagogy. It should be noted that we counted on ordinary support from the administration and expected the students to follow up the coursework in a responsible way: *1) Take calculated risks to cultivate novelty and creativity.* In order to test what the technology can do, there must be as few rules as possible, and the risk involved in operating the drone for such exploratory purposes must be handled out of the box. *2) Teach the students to be reflective about the constraints of the technology.* In the course of their exploration, the students were supposed to evaluate the rules and potentially acknowledge that the constraints on exploration were rational and necessary. For the teacher, this approach requires the ability to stimulate problem-solving processes. *3) Provide a structure for the students' work with time-boxing of iterations and deliverables.* In order for creativity to blossom, there must be strict but empty limits to it. The teachers should try to create suitable cycles of production sprints and evaluation sessions that lead toward a final delivery. Such procedures may be established as routines in the educational program in the future.

From January to June 2016 students underwent practical training in drone flying and safety instructions, including the use of an operation manual. The students were divided into groups to design low-fidelity prototypes in an iterative process. There were four workshops, with evaluations of the prototypes by a panel of researchers (the authors of this article). The groups produced prototypes with a low technological readiness level but with validated journalistic potential. Specifically, the groups were charged with designing a low-fidelity prototype of an app for the 3DR Solo drone and producing a one-minute pitch video about the app's intended functionality.

We interviewed 12 people when the course had been completed. There were seven students, three teachers and two administrators, and they were interviewed by the three authors during a period of one week. The interviews dealt with 1) flying, 2) rules, 3) programming, 4) journalism and 5) course evaluation. There are methodical challenges to interviewing almost everybody involved in a course. Firstly, all our informants were fully aware of the experiment before we interviewed them and therefore had a high level of meta-knowledge and were strikingly reflective about their experiences. Secondly, most of the informants knew each other and could potentially have learned the identity of the others. Thirdly, the interviewers were involved throughout the course and were known by the informants as stakeholders in the course. We are acutely aware of these issues and counteracted them by paraphrasing some of the informants' statements and by being open about the potential conflict of interest. Please note that the interview guide and consent form were approved by the Norwegian Data Inspectorate. The interviews were transcribed, and the analysis is based on pen-and-paper readings of the material.

Analysis: three degrees of risk taking

In this section, responses from informants are presented and interpreted, with the students first, then the teachers and finally the administrators. The analysis gives equal weight to these different functional roles in the course.

1: Risk stimulates creativity among students

Students experienced that being allowed to fly an actual drone was a type of risk that stimulated their motivation to test the technology further. To feel what it was like to fly a camera drone was better done outdoors than indoors. One student says: "Flying the drone outdoors was scary, but mostly fun. I liked the adrenaline kick that you get because there are so many things that can happen." In contrast, flying the small drones manually indoors, with or without cameras, was harder than expected for the students. These drones proved difficult to keep under control, and frustration made students lose energy and interest. Outdoors, students felt that they were in control with the flying camera and were challenged not by the technology itself but primarily by uncontrollable weather conditions. In particular, strong wind was a risk factor that made the participants reflect on possible loss of control and its consequences. One student says that he could "feel the sense of losing control as the drone went higher. It disappeared behind me, over an area with parked cars. It was scary. I don't want to damage cars. Not fun!"

The ten-minute slots that students were allowed to test the 3DSolo drone outdoors came to an end abruptly after a few weeks. Since students were not allowed by the administration to practice drone flying as much as they wanted, there was a period when the administrative rules were broken. A month into the semester, one of the teachers allowed students to fly the 3DR Solo drone informally in a nearby park. One student describes the event where a teacher broke the rules to support the students' piloting experience: "He said that we should go out there and try and make mistakes, that it was ok. We had a flight each and tested the various functions so that we knew how they worked. We could start to think about what was already implemented and what could be developed further, and what we could make out of it." Other students found the rule-breaking OK as well as long as the teacher was watching what was going on. One student said: "Since everybody understands that the drone can land on a car or something, there is a tension between what you dare to do and all the creative stuff that might come out of it."

After a few weeks of drone flying, the administration found out and decided to confiscate the drone due to concerns about lack of insurance. If anything happened, the university would most probably be liable for the full cost. The students had to wait for several weeks before they were allowed to fly the drone again, and then only for five to ten minutes under supervision of the drone teacher. This decision was disappointing to everyone, and it deflated students' creative energy. "We would have liked to fly more with the big drone, but we know why we couldn't," one student says obediently. The confiscation influenced exam materials because the drone teacher had to produce aerial footage on behalf of the students. Students had good ideas that could not be continued. One student says: "We had high expectations, because of all the cool videos on YouTube and the Internet. And you think, "Ah, we must do this too!," but it turns out you can't. The drone teacher had to do it all, and he didn't do it the same way I would have wanted."

Students experienced that flying the drone outdoors in safe areas and in accordance with aviation regulations served as constructive risk taking as long as they felt they were in control of the drone. Since the teachers were perceived to be in charge of the flying, also during the period when the administrative rules were broken, students felt they were taking a lower risk than what seemed to be the case with teachers and administrators. One student comments: "The teacher is experienced, and was thinking about the consequences while we were only thinking about possibilities. For example we wanted more spectacular images from the sea, but the end result was a little more boring than we had thought." Students had a more risk-taking attitude than the drone teacher. The question that several students posed was whether more flying would have made them more careful by the end of the course: "Maybe if we had been allowed to fly more, we might have thought it over more too, or something would go wrong," one student says.

At the end of the semester, the students submitted prototypes of drone apps for journalism, and their deliverables showed a real engagement in how drones could be designed to benefit the public sphere. One student explains his/her motivation like this: "We thought about the requirement that it should be useful for a journalist, and we saw that there were no particular products in that sector. So, we decided on creating the drone-rig, where a TV-journalist can go out and do a live report alone if there is no camera person available. You will not need to go to a course or take an education in drone flight to be able to use the drone rig; you just grab it and run." In relation to the rules for drone piloting, this is a radical proposal because rules specify that a drone cannot be used without first notifying everybody in its vicinity during shooting and engaging a team of at least two or three persons. Since a TV station would presumably want to be the first with a breaking news story, these rules would presumably not be followed, or they would make the concept void if they were followed.

2: Risk stimulates calculated risk taking among teachers

The three main teachers in the course are creative experts in their fields: media design, drone piloting and programming; their role was to enthuse and engage students in learning the skills selected for our purposes. Teachers have a duty to support a culture of fair and transparent testing criteria. The course leader was formally responsible for making a fair assessment of the deliveries, considering that the grades will appear on the students' CVs afterwards.

It is an intense experience to teach somebody to fly a vehicle. One teacher says: "The first time I saw that you could bring a camera up into the sky I just decided to have it, and I started building it myself. And when I flew it I realized that "Wow," this is not only a possibility to bring something up to the sky, but it is also fun." The teachers are acutely aware of the powers that this "fun thing" has. One of the teachers says: "You must be vigilant when you are doing this, it is not a completely streamlined user experience. You cannot throw the drone up and lower your shoulders; you must watch out all the time. There is a real risk of injuring people, and it is important that the pilot and the assisting team know what they are doing." The other teacher also talks about the risk of hurting people. "Whenever people call to discuss an assignment, I ask about the surroundings so that I know what risks are involved. If something goes wrong with the device and it starts falling down – what then? The worst-case scenario is that it hits somebody in the head, in the hand or body, because this is the most important thing in this life – people, right?" While the teachers recommend that students should fly actively, they were concerned about managing the risks beforehand. One teacher says: "To get nice shots sometimes requires taking a risk. Definitely. But then you validate these risks, and you are prepared. If something happens my fingers will move that way, to make my planned exit from this risk. As a human being controlling the device I am prepared for that, and I pay attention to all the movements of the device itself, the wind movements, sunlight, people's movements."

There is a difference among teachers in their willingness to take risks. One of the teachers allowed the students to test the drones without prior consent from the administration. Why did he do this? Interestingly, he said it is important to teach the students to have "copper in the attitude," meaning that you have to dare to push some boundaries. "If one is too squeamish out of the starting blocks, one gets nowhere. You need to speed up, fire all the guns, and then be careful too, and not do this at a public square, but on a grassy field." The teacher is concerned that the students should manipulate the technology directly: "In order to teach them something, we must have access to the metal, and we can connect straight to the drone CPU with a cable and do what we want. And for the learning process nothing is better than just exploring, with full access, and test out everything, and get to know how it fits together." This teacher's attitude cultivates independent decision making among students. He would allow any dedicated student to test and program the drone in whatever way the student found worthwhile, and make it his job to "reduce the anxiety that the students felt in relation to the risks."

3: Risk stimulates carefulness among administrators

Our interviews with administrators add a critical and new dimension to the understanding of journalism and media education. We expose decision-making hierarchies that cause tensions when high-risk technologies are employed.

Whereas risk management procedures are integrated in the management of disciplines, such as engineering and chemistry, social science departments are traditionally less experienced in managing risk related to technological experiments. Our informants work in a department in which risk management of this kind is not the norm. Regular media and information science teaching and research do not involve explosions, human injury or rule-breaking. On the contrary, the social sciences and humanities are a rather quiet branch of higher education where management colleagues are likely to be upset at suddenly being asked to take risks involving drones and students. Most interestingly in this case, it was not the physical danger of flying drones that dominated management's rhetoric; it was the fear of breaking aviation and insurance rules and getting into trouble – regardless of whether or not there were any accidents (in our situation, there weren't any accidents.).

Risk induces administrative accountability and a sense of responsibility in university management. The rules for drone flying issued by the aviation authorities will, by themselves, require university administrators to behave carefully. In relation to students and teachers, the role of management is to ensure that rules and regulations are followed when experiments (such as drone flying) are carried out within the frames of a university course. An administrator explained, "Since there is so much fuss about this now, in my opinion the university should make a legal assessment of how to tackle the situation. The university's lawyers always have to think about the "worst case scenario," and they would rather say no than yes. They are afraid that they will give advice that turns out to have a boomerang effect. So we would have to prepare the case well and follow the regulations from the aviation authorities.

There is a willingness to find solutions but also a pragmatic realization that things take time. The administrative staff agree that there are certainly going to be better solutions next year. There was interest in sharing accountability across the Higher Education Institution (HEI) landscape and a willingness to learn from other university sectors, for example, archaeology and geography, where researchers fly drones for research purposes. The informants displayed interest in collaboration with private companies and other state institutions such as the NRK, the Norwegian Broadcasting Corporation. These external institutions take an interest in the university's drone education. "NRK signalled that they are very interested in establishing a university drone pilot certification to recruit from." The NRK drone pilots and television producers have much experience, and the administrative staff listen to their advice. For example, there are several classes of drones with different certification requirements. The university flies the smallest class (RO1). However, one informant tells us that "The NRK people recommended to us to go up one level. There is so little you can do with the smallest type, regarding where you can fly and things like that. If you go up a class, you can fly more actively and ask permission to fly in a city and other locations that you would otherwise not be allowed to." Here, the ambitions of the NRK are followed and included in the desired educational quality of the higher education institution.

Despite the creative gain achieved with high-risk behavior, it is clear that higher education institutions have to comply with the law. Administrators, backed by law consultants, have legitimate reasons to work with worst-case scenarios for what can happen during teaching sessions. Teachers and students must also accept the fact that it takes time to find safe and rule-compliant solutions to the problems posed by introducing new high-risk technologies in a higher education institution.

Conclusion: four learning principles

While it was interesting to see how many actors were involved in drone management, the students were always the main protagonists. We exposed the students to drone technology in practice and challenged them to explore its potential for journalism in a process called responsible innovation pedagogy. What worked and what did not work so well? We wanted to arrange a course not only to learn about the creative potential of drones (and not only to learn about the immanent physical risks and safety regulations associated with drone flying) but also to stimulate the students to reflect on and be critical about how such a technology can be applied in journalism in a responsible way – according to the norms of news journalism. It turned out that the parties involved were concerned about the risk-taking aspect of the drone technology. The perceived risks of drones made administrators and teachers act so carefully that the students' creative process was slowed down, and learning was involuntarily reoriented to rule-following. The general impression was that everybody sought a balance between the perceived risk on the one hand and their shared responsibility on the other. However, the general impression was also that the individual's perception of safety in the situation was the most dominant influence on the learning process. Thus, risk taking was less prevalent than attempts to gain more control.

When interpreting the behavior reported in our analysis, we saw that the learning curve was significant for the students, teachers and administrators, albeit differentiated. In this process, students learned about risk directly; high-risk sensations appeared to stimulate learning and ownership of the product. Teachers attempted various versions of calculated risk, where students were given leeway but under a watchful eye. Ideally, teachers should teach drone flying in the same way that a chemistry teacher does when creating an explosion in the classroom. Students learn something about the forces with which they are dealing, but they do so in a controlled situation where the level of possible damage is limited. Our design experiment was, however, not based on established knowledge about how to teach drone flying. As such, it involved greater risks for everyone. While this risk can be considered a weakness in the course planning, it also showed that teaching drones in media education is an emerging practice. There is little or no prior competence.

The validated insights from the design experiment previously described lead us to formulate four learning principles that a higher education institution course should have in order to constitute a responsible way of teaching students to use high-risk technology:

- Focus on *context understanding* in the practical setting where the high-risk technology is going to be used. Make field trips and walk the terrain. Students must acquire the necessary motor skills to deal with the practical situation of using the high-risk technology.

- Allow for *independent decision making* regarding the creation of drone-based media products. Students should make as many decisions about aesthetics and content as possible, limited mainly by laws and regulations. Teachers should make as few decisions about creative direction as possible.
- Cultivate the strategic ability to *anticipate* any implications of their high-risk technology prototypes for media innovation in the future. In functional terms, what sector of an industry would likely adopt this new technology and its practices?
- A *level of reflexivity* sensitive enough to address the big question of what is at stake for society in general in relation to technology development is needed. Students should be imaginative enough to consider the good and bad consequences of new technologies before they are made and promoted in society.

These learning principles should ideally characterize any exploration of high-risk technology in higher education institutions.

References

Antonsen, S. (2009) *Safety Culture: Theory, Method and Improvement*. London: Ashgate.

Barab, S. and Squire, K. (2004) Design-Based Research: Putting a Stake in the Ground. *Journal of the Learning Sciences*, [online] 13 (1), pp. 1–14. doi: http://doi.org/10.1207/s15327809jls1301_1.

Brown, A. L. (1992) Design Experiments: Theoretical and Methodological Challenges in Creating Complex Interventions in Classroom Settings. *Journal of the Learning Sciences*, [online] 2 (2), pp. 141–178. doi: http://doi.org/10.1207/s15327809jls0202_2.

Collins, A., Joseph, D. and Bielaczyc, K. (2004) Design Research: Theoretical and Methodological Issues. *Journal of the Learning Sciences*, [online] 13 (1), pp. 15–42. doi: http://dx.doi.org/10.1207/s15327809jls1301_2.

Darsø, L. (2011) *Innovationspædagogik – kunsten at fremelske innovationskompetence*. Copenhagen: Samfundsliteratur.

HEIRRI Higher Education Institutions and Responsible Research and Innovation. (2017) *Front Page*, [online]. Available at: http://heirri.eu [Accessed 13 Sept. 2017].

Hmelo-Silver, C. (2004) Problem-Based Learning: What and How Do Students Learn? *Educational Psychology Review*, 16 (3), pp. 235–266.

Hollnagel, E. (2014) Is safety a subject for science? *Safety Science*, 67, pp. 21–24.

Kettunen, J. (2011) Innovation Pedagogy for Universities of Applied Sciences. *Creative Education*, 2 (1), pp. 56–62.

RRI Tools (2017) Welcome to the RRI Toolkit. *Front Page*, [online]. Available at: www.rri-tools.eu [Accessed 13 Sept. 2017].

Säljö, R. (2010) Digital Tools and Challenges to Institutional Traditions of Learning: Technologies, Social Memory and the Performative Nature of Learning. *Journal of Computer Assisted Learning*, [online] 26 (1), pp. 53–64. doi: http://doi.org/10.1111/j.1365-2729.2009.00341.x.

von Schomberg, R. (2012) Prospects for Technology Assessment in a Framework of Responsible Research and Innovation. In M. Dusseldorp and R. Beecroft (eds.). Technikfolgen abschätzenlLehren: Bildungspotenziale transdisziplinärer Methoden. Wiesbaden: VS Verlag für Sozialwissenschaften.

Sternberg, R. J. and Lubart, T. I. (1999) The Concept of Creativity: Prospects and Paradigms. In R. J. Sternberg (ed.). *Handbook of Creativity*. pp. 3–16. London: Cambridge University Press.

Stilgoe, J., Owen, R. and Macnaghten, P. (2013) Developing a Framework for Responsible Innovation. *Research Policy*, 42 (9), pp. 1568–1580.

Strand, R. (2015) Indicators for Promoting and Monitoring Responsible Research and Innovation. *European Commission*, [pdf]. Available at: http://ec.europa.eu/research/swafs/pdf/pub_rri/rri_indicators_final_version.pdf [Accessed 13 Sept. 2017].

Weilenmann, A., Säljö, R. and Engström, A. (2013) Mobile Video Literacy: Negotiating the Use of a New Visual Technology. *Personal and Ubiquitous Computing*, [online] 18 (3), pp. 737–752. doi: https://doi.org/10.1007/s00779-013-0703-x.

Wold, T. (2016) *Procedures coming every day: Safety Management Systems and safety communication in high-risk industries*. PhD. Norwegian University of Technology.

7 Three scenarios of responsible drone journalism

Turo Uskali and Astrid Gynnild

Introduction

Since the turn of the millennium, scenario development has become a new buzzword, particularly in business and governance. Therefore, we begin this chapter by crushing a widespread myth: Scenarios are not predictions (Van der Heijden et al., 2002). Scenarios are made when predictions are meaningless or out of reach. Subsequently, we suggest that a scenario might be considered a coherently structured speculation (van Notten, 2006) on aspects of a phenomenon based on a variety of accessible data.

Scenario building has proved useful to start debates on future society, especially when we otherwise might be overwhelmed and scared by the many uncertainties in a field.

In his book on surviving the techstorm, Nicklas Bergman (2015) wrote that *planning for uncertainty* is "not about trying to guess the one outcome that will occur" (p. 179). Rather, he pointed out, planning for uncertainty means being prepared for the multiple possible outcomes of an issue while trying to understand the most likely outcome. We think that is a good way of looking at this.

On the Internet, there is a jungle of websites specializing in scenario building. The examples and steps provided may at first glance look fascinatingly simple and convincingly clarifying. But by choosing such an approach for this last chapter, we of course do run the risk of simplifying complex issues that go far beyond journalism. One of the things we discovered was that scenario development requires from facilitators that the purpose and tools for the process be explicitly decided beforehand; a scenario will only be good as far as it goes.

Simultaneously, the scenario approach aligns with the basic principles of responsible research and innovation (RRI) that were introduced in Chapter 1 and that form the basis of this research project. Future orientation and foresight are strongly built into the responsible innovation paradigm. Researchers and stakeholders are encouraged to engage actively in ongoing

DOI: 10.4324/9781315163659-7

developmental processes, for instance in visual surveillance technologies in the making. They are invited to raise public awareness in a number of ways, in which the first and foremost task is "to ask what futures do we collectively want science and innovation to bring about, and on what values are these based" (Owen et al., 2013, 37). Thus, we believe that in order to answer what responsible drone journalism is in an RRI-context of anticipation, reflection, deliberation and responsiveness, scenario development is a good place to start the broader discussion – and to finalize this book with an open ending.

In the following, we will first discuss opportunities and dilemmas of scenario development as an exploratory approach. Next we will sketch three simple scenarios of responsible drone journalism and finally, we will suggest ways that readers of this chapter might engage in developing scenarios that are meaningful and relevant for them – in their particular local contexts.

Further definitions of scenario development

Historically, scenario development emerged from strategic planning in the military after the Second World War. The concept now refers to a variety of approaches by businesses, governance, foresight studies and participatory future initiatives. When the European Commission (2017) used scenario development for mapping the future for research and innovation policies in Europe, the report described only two possible outcomes: negative – the trends go on unmanaged and uncontrolled, or positive – society takes action. We hope there are more ways to go within one or the other of these outcomes. Normally, prominent future researchers prefer to provide at least four to five scenarios to play with. Scenarios are intellectual tools to help imagine a *variety* of future trajectories, and as van Notten (2006) points out:

> Scenarios are consistent and coherent descriptions of alternative hypothetical futures that reflect different perspectives on past, present, and future developments, which can serve as a basis for action.

In his comprehensive review of scenario characteristics, van Notten differentiates between educative scenarios, in which exploration and awareness raising is a main aim, and scenarios as decision support or as pre-policy research. In practice, he concludes, scenarios are typically hybrids of explorative approaches and pre-policy research. They are often conducted in two steps; exploratory approaches usually provide a necessary first overview, but tend to be too general for decision making.

Van Notten also distinguishes between process-oriented scenario development to promote learning and communication skills, and product-oriented scenarios, which focus more on the end products and less on the processes.

Also, scenario characteristics might be split into goals, design and content (van Notten, 2006).

That is, they have capability to forecast dilemmas and opportunities based on vague data. We know from technology history that in general, it is very challenging to foresee the roles or successes of new devices in future communication. For example, ubiquitous technologies such as smartphones and the Internet were initially developed only for special purposes; the Internet for scholars (1969) and smartphones for business people (1993, Nokia Communicator).

Emergent technologies, of which drones are a typical example, often start in the military and are later sold to the civilian mass market. No wonder there often exists a long-lasting anxiety and suspicion toward new technologies among large groups of people. In literature, this phenomenon and natural attitude is often defined as techno-pessimism. At the other end of the same continuum we find the techno-optimists (Thierer, 2010). Techno-optimists, in this case drone optimists, are those who might benefit the most from the new technology, like manufacturers, sellers and "heavy-users," often hobbyists. In addition, law enforcement officials, architects, property brokers, firemen, rescue workers and journalists are among the professional groups that take advantage of drone technology in their daily work.

Drone footage is a ubiquitous element of artistic storytelling in the fiction side of media productions like movies, TV series and other forms of entertainment. You just cannot avoid drone footage when watching entertainment. And still we know fairly little about the further adoption and adaption of drones by news media and how, for instance, drone traffic, in general, might be regulated in the near and distant future. Therefore, we wanted to play the role of futurists on these last pages of the book.

The scenarios we sketch out are simple in several respects: They are based on weak signals or early warnings that emerged from empirical data (Ansoff, 1975; Ansoff, 1980). The weak signals mostly originated from online news sources and related to, for instance, photo competitions and awards, interactive map and app development, educative events, new legislative rules in the United States and upcoming EU rules as well as discussions on nano-drones and new threats by hostile drone environments.

We chose to focus primarily on one contingency for the further spread of responsible drone journalism, namely, the use of bans as a legislative means. The threat of bans is currently at stake in many countries, and the threat is real; governmental bans are issued with varying implications. For shorter or longer periods of time, total bans have been implemented in authoritarian countries as well as in a social democracy such as Sweden (see Chapter 2).

The drafted scenarios are simple also in the sense that they are not based on focus group discussions, public debates or other staged events but on written data collected for the study. Van Notten's typology proved to be particularly applicable to the scenarios and helped contextualize the exercise.

First scenario: drones everywhere

Based on existing empirical data on drones, our first scenario is that drones will become ubiquitous everyday tools. In journalism, this means that aerial imagery of everyday situations and events will be the new normal. An early indication of the upcoming normality of drone footage in journalism is found by looking at recent international photojournalism awards. For example, the *New York Times* photographer Josh Haner was awarded Pictures of the Year International Awards in 2017 for his documentary project on climate change. Haner has used camera drones in his work since 2012 (Estrin, 2017). The same year, the photographer Sami Kero from the *Helsingin Sanomat* won thc prestigious Daily Press Awards at the International Photojournalism Festival in Perpignan. His winning series of Finnish ice hole swimmers was captured by a camera drone (Koppinen, 2017).

Awards in professional photo competitions are prestigious markers of new trends in journalism, and the symbolic power of a jury's decisions should not be overlooked. The use of drones for journalistic purposes spread from entrepreneurs and innovative newsrooms to the rest of the news industry. Haner predicts that especially the miniaturization of drone technology brings drone journalism as a newsgathering tool one step further. Without hesitation, he claims that,

> [j]ust like the influx of digital cameras and camera phones created a saturation of imagery, we're going to have to adapt and figure out how we can bring our creativity to these new technologies. – I think it's only a matter of time before we have micro drones with high-quality cameras that reporters can take into the field.
>
> (Estrin, 2017)

The increasing demand for drone journalism education is another clear signal in support of the *drones everywhere* scenario. As the first drone journalists from 2011 to 2015 typically operated mostly in the new online cultures of learning and were self-taught, the evolving next wave of drones in journalism, from 2015 on, have created a demand for more systematic education in drone journalism. Many local newsrooms invest in drone boot camps aimed at efficient drone learning for their journalists and photographers. These are signs of responsible drone journalism, as are new licensing requirements for professional drone operators in many countries.

From the beginning, dronalism has benefited from hobbyists and activists who have provided drone footage to the newsrooms. This model of networked journalism (Beckett, 2008), in which amateurs and professionals cooperate, plays an important part of this first scenario. Even if we should be critical toward marketing messages such as "eventually drones will be in every household" for avoiding the hype-effect, the number of drones sold for civilian purposes does grow exponentially. The increasing flow of user-generated content is well known from ubiquitous smartphones, even though user-generated content is rarely mentioned by news organizations (Wardle et al., 2014). This could also be the case with user-generated drone footage. We see more aerial imagery, but photographer bylines are typically missing.

The *drones everywhere* scenario is based on the assumption that legal frameworks will be developed to ensure that the drones will be available for the citizens and entrepreneurs to innovate new drone-related practices and services. The scenario consists of only modest regulation, which will be regularly updated and changed by the politicians based on feedback from all parties involved, including drone operators and authorities as well as civilians.

As pointed out in previous chapters, the legal framework for drones is still under construction in different parts of the world. But just by monitoring the situation in Europe and in the United States, it appears that in general, politicians are very positive to the potential of drones.

In the United States, the FAA in 2016 opened up for civilian drones in many businesses, including journalism. This legislative change was meant to ensure that guidelines for the responsible use of the drones were followed.

In similar vein, the European Union aims at harmonizing drone regulation in Europe. The EU plans to introduce new drone rules in 2018, and according to the new EU rule drafts, there will be three different categories for drone activities: open, specific and certified. The open category is mainly for hobbyists who will not need any flying permits. The second category could include, for example, the aerial packet transportation of service providers. The third, certified category is created for heavy-weight drone operations such as drone taxi services. This category consists of detailed instructions and rules, and special permits and risk assessments are needed. Hearings of the upcoming EU drone rules (Helsinki 22.8.2017) indicate that there will be a transfer time, at least until 2021, for all the member countries to adjust the new rules to national legislations.

Moreover, the European Union plans to create a special airspace for drones called U-space. This lower airspace will develop its own traffic management systems (Ec.europa.eu, 2016), and the services of the new U-space are planned to begin around 2021, with full services available tentatively in 2030.

Interestingly, according to the new EU rule drafts, drone piloting will require more systematic education in the future, especially when flying over 50 meters. Online tests are developed, and probably also practical tests in certified institutions. The highest possible altitude for the drones will be 120 meters. In the near future, the EU will also require that drone manufacturers and retailers become responsible for constantly updating the geofencing and safety features of the drones that they sell. This first scenario, *drones everywhere*, is indeed the most positive one and might also be also called the "Drone Age" (Economist, 26 September 2015).

Second scenario: total ban of camera drones

The second scenario is based on the assumption that all the positive developments of drone journalism will be interrupted by one or more fatal drone incidents. This scenario is named the *total ban of camera drones* and is a so-called worst-case scenario.

Only one devastating incident might be enough to ground all current drone activities at a local or national level. So far, thousands of incidents in the air caused by migrating and other birds are reported (Wired.co.uk, 2017), and some of these cases cause serious problems for airplanes. For example, a US Airways plane had to execute an emergency landing into the Hudson river in New York because it was hit by a bird on takeoff from LaGuardia airport (CNN.com, 2009). If the bird had been a drone, the consequences might have been fatal.

Great dangers lurk, especially if drones are operated near airfields or at high altitudes, close to commercial airline routes. Globally, there are already almost 100 reports that document the threats to airplanes by drones, mainly caused by a few overenthusiastic hobbyists. No one knows yet precisely what kind of consequences a crashing drone would cause, but one can foresee serious problems, especially if the drone smashed into the turbines (Wired.co.uk, 2017).

Even before any fatal drone incidents, many countries have banned the use of drones as a preventive measure – from authoritarian Nepal to liberal democratic Sweden. In 2016, Sweden temporarily banned the use of camera drones after intelligence authorities warned that drones might be used in terrorist operations in the country. The local drone industry, and also media organizations, accused the government of harming their business interests with the ban, which was based on century-old legislation. In the summer of 2017, the Swedish government announced new rules for the use of drones and ended the total ban. However, the Swedish case demonstrates the ease with which a total ban might be implemented, if needed.

Another threat often disseminated by the new media is the use of drones for terrorist attacks. The *New York Times* (September 23, 2017) published evidence that, for example, The Islamic State, ISIS, used small consumer drones for their warfare in Iraq and Syria. US military sources argue that these "airborne improvised explosive devices" are causing a global threat. Furthermore, according to the story, in the United States "the authorities voice increasing concerns about possible Islamic State-inspired drone attacks against dams, nuclear power plants and other critical infrastructure" (Schmitt, 2017).

Third scenario: drone mosaic

Our third scenario, the *drone mosaic*, is taking the middle path between the two scenarios already explored here. Drone regulations vary from country to country, and even year by year. An interactive world map developed by a blogger and traveler provides indications of the full spectrum of operative drone regulations, from total bans to only modest regulation. (Polat, 2017; Simpson, 2017). This data visualization was published with the help of Google maps and is also available as a smartphone app. The map, split into four color-coded categories, should be critically consumed, as users frequently detect new errors. But it does serve a function as the first rough draft of a global drone mosaic in the making. As of September 22, 2017, the map indicates that there are 74 green countries from United Arab Emirates to Puerto Rico, and 40 yellow countries from Austria and Belgium to Vietnam and Vanuatu. Green means that "drone use is generally allowed," and yellow that "drone use is limited or may require cumbersome registration processes."

Perhaps most interestingly, 40 countries have declared "the total ban" or are "heavily restricted" for the use of camera drones. These "red" countries include a wide spectrum of states and areas, starting from Antarctica, Bangladesh, Bahrain and Brunei, and ending with Vatican City and Venezuela. Altogether, 85 countries or areas were categorized as "gray," meaning "no data" is available or there are "no defined or applicable UAV laws in the country." Many of those "gray" countries are in Asia, Africa or South America, but also South Korea was in the same category. Indicatively, neighboring countries could have totally opposing drone rules, like "green" Arab Emirates and "red" Oman and Qatar (Google Maps, 2017). Based on Polat's map, more than 100 countries allow the use of camera drones, which indicates that drone journalism has large testing fields for aerial newsgathering and storytelling. It also indicates that responsible drone journalism as a newsbeat is indeed needed on all continents.

Most likely there will never be a legally unified global approach to the use of military and civilian drones. This leads us to suggest that a mosaic model might be the most probable scenario for the future; some countries and areas in the world will spearhead and further develop drones for a multitude of operations in society. Local drone hubs and industry clusters are built in many places outside of China, which is currently the leading manufacturer of consumer drones. In some areas and countries, the development or use of drones will not be allowed at all.

Satellites next

So far in this book we have mostly focused on opportunities and dilemmas of drones as a visual newsgathering tool in journalism. The idea was that in order to gain competence in covering the drone field as a news beat it is helpful to explore the options of a new technology from the inside out. In the first chapter, we also proposed that at this point in time, there is a crucial opening for journalists to explore, inform, influence and impact the further direction and governance of drones in society. New features are constantly added to drone technology, features that expand the range and sensoring conditions of its operations. We believe it is time for responsible drone journalism to expand the scope; it is time to integrate the exploratory camera drone experiences with critical thinking and constructive solution-foci on behalf of society as a whole.

As devoted news hunters, entrepreneurial drone journalists are strategically placed in the middle of merging interactive communities that posit more communicative power than they might be aware of. As members of global networks, drone journalists tap into the creation of global marketplaces and new connected cultures of learning in which the pace of good ideas escalates exponentially. With increasing time pressure comes increasing quests for action. Thus a crucial question is in what ways might journalism contribute to make individuals, governments and populations more insightful decision makers of an expanding technology such as drones. How can open collaboration communities and the free exchange of ideas best be used for the common good of society? In what ways can journalism programs, as labs for journalism innovation and exploration, merge the dualities of responsible drone journalism? And in what ways might scenario building for a responsible future be incorporated in higher education as well as in virtual learning spaces?

The increasing time tension that follows from digital connectivism leaves less time for reflection. Simultaneously, as pointed out by Castells (2012), the Internet as a social network provides spaces of autonomy beyond the control of governments and corporations that previously monopolized communication power. Exactly at the nexus of transformative transparency and surveillance, shifting legislation and virtual communication spaces lie the new options for ubiquitous learning. We suggest that we are entering a new era of learning in which we will see amazing progress in educative approaches, formally and informally. We believe that the framework of responsible research and innovation (RRI) (Owen et al., 2013) will be a valuable tool in this transition. The website on RRI tools is a good place to start for those who want to further engage students, colleagues, politicians and others in face-to-face debates on the future of drones in society. Playing the infinite game of technology (Kelly, 2009) from a human perspective means not to focus on details, first and foremost, but to be value-driven, to have a vision not only of what technology wants but what we as humans want.

And while writing these lines, technology is doing for us what humans have always envisioned – and feared. It is time to look higher up into the sky than lower airspace. Far out in the endless horizon, beyond our human gaze, are the satellites. Similar to drones, satellites were invented as spying devices in the 1950s. When the Soviets started the space race with sputniks, the Americans followed. And later, from the 1980s onward, satellites have enabled many other commercial global communication systems. Foreign reporting, for instance, has been enabled by satellite phones with video feed points and links for decades.

With nanotechnology, miniaturization is expected to be the next big trend. Small satellites are now called toasters. They are still quite expensive, but there are services available to individuals and organizations on a global market. The miniaturization of the satellites has opened new markets for innovative companies operating hundreds of "mini-satellites." Their services include, for example, real-time Earth monitoring, and satellite photo archives. Both might be useful as new data sources for the news media, and many governments offer satellite imagery services. ProPublica was among the first news media to use governmental satellite imagery for their investigative reporting (ProPublica.com, 2014).

With the advancing drone and satellite technologies, the concept of eyes in the skies has taken on a new meaning. We propose that satellite journalism might become the next test of responsible research and innovation in future society.

References

Ansoff, I. 1975. Managing Strategic Surprise by Response to Weak Signals, *California Management Review*, vol. 18, no. 2, pp. 21–33.

Ansoff, I. 1980. Strategic Issue Management, *Strategic Management Journal*, vol. 1, pp. 131–148.

Beckett, C. 2008. *SuperMedia: Saving Journalism So It Can Save the World*. London: Wiley-Blackwell.

Bergman, N. 2015. *Surviving the tech storm*. London: LID Publishing.

CNN.com. 2009. Airplane Crash-Lands into Hudson River: All Aboard Reported Safe. http://edition.cnn.com/2016/08/11/us/hudson-landing-archive-news-story/index.html [Accessed 10 Sept. 2017].

Ec.europa.eu. 2016. https://ec.europa.eu/transport/modes/air/news/2016-11-23-drones-commissioner-bulc-presents-plans-creation-european-drone-services_en [Accessed 10 Sept. 2017].

Estrin, J. 2017. Democratizing the Sky: Drones in Visual Journalism. *Lens.blogs.nytimes.com*. https://lens.blogs.nytimes.com/2017/03/16/democratizing-the-sky-drones-in-visual-journalism/?emc=edit_tnt_20170316&nlid=23071106&tntemail0=y [Accessed 27 Sept. 2017].

European Commission. 2017. New Horizons: Future Scenarios for Research & Innovation Policies in Europe. https://publications.europa.eu/en/publication-detail/-/publication/b2d78a84-3aae-11e7-a08e-01aa75ed71a1/language-en [Accessed 18 Sept. 2017].

Google Maps. 2017. Drone Laws for Every Country. www.google.com/maps/d/u/0/viewer?mid=1OkEtyCaGNjKhLeMr6L2IU975SP8&ll=21.19352635444651%2C55.58338756250009&z=6 [Accessed 22 Sept. 2017].

Kelly, K. 2009. *What technology wants*. New York: Penguin Books.

Koppinen, M. 2017. Helsingin Sanomien valokuvaaja Sami Kero voitti avantokuvillaan suuren kuvajournalismin kilpailun. [The Photographer of Helsingin Sanomat Won a Prestige Photojournalism Award]. www.hs.fi/kulttuuri/art-2000005357361.html?share=e63ae6922445ca585fa14ada466defb4 [Accessed 27 Sept. 2017].

van Notten, P. 2006. Scenario Development: A Typology of Approaches. In *Think Scenarios, Rethink Education*. OECD Publishing: Paris. http://dx.doi.org/10.1787/9789264023642-6-en.

Owen, R., Bessant, J., and Heintz, M. 2013. A Framework for Responsible Innovation. In Owen, R., Stilgoe, J., Macnaghten, P., Gorman, M., Fisher, E., and Guston, D. (Eds.) *In Responsible Innovation: Managing the Responsible Emergence of Science and Innovation in Society*. Chichester, UK: John Wiley & Sons, Ltd. doi: 10.1002/9781118551424.ch2.

Polat, A. 2017. This Map Shows You the Drone Laws for Every Country in the World (Updated Regularly). 25.7.2017. *Foxnomad.com*. https://foxnomad.com/2017/07/25/map-shows-drone-laws-every-country-world-updated-regularly/ [Accessed 22 Sept. 2017].

ProPublica.com. 2014. http://projects.propublica.org/louisiana/ [Accessed 10 Sept. 2017].

Schmitt, E. 2017. Pentagon Tests Lasers and Nets to Combat a Vexing Foe: ISIS Drones. *Nytimes.com*. www.nytimes.com/2017/09/23/world/middleeast/isis-drones-pentagon-experiments.html?mcubz=1&_r=0 [Accessed 27 Sept. 2017].

Simpson, J. 2017. Here's a Map with Up-to-Date Drone Laws for Every Country. *Petapixl.com*. https://petapixel.com/2017/09/20/heres-map-date-drone-laws-every-country/ [Accessed 22 Sept. 2017].

Thierer, A. 2010. The Case for Internet Optimism. In Szoka, B., and Marcus, A. (Eds.) *The Next Digital Decade: Essays on the Future of the Internet*. Washington, DC: TechFreedom.

Van der Heijden, K., Bradfield, R., Burt, G., Cairns, G., and George Wright, G. 2002. *The Sixth Sense: Accelerating Organizational Learning with Scenarios*. Chichester, UK: Wiley & Sons.

Wardle, C., Dubberley, S., and Brown, P. D. 2014. Amateur Footage: A Global Study of User-Generated Content. In *Tow Center for Digital Journalism*. New York: Columbia University Press. https://academiccommons.columbia.edu/catalog/ac:2v6wwpzgnj.

Wired.co.uk. 2017. Silicon Valley Sexism, Drone Drama: Podcast 325. www.wired.co.uk/article/podcast-325 [Accessed 27 Sept. 2017].

Index

For Product Safety Concerns and Information please contact our EU representative GPSR@taylorandfrancis.com
Taylor & Francis Verlag GmbH, Kaufingerstraße 24, 80331 München, Germany

www.ingramcontent.com/pod-product-compliance
Ingram Content Group UK Ltd.
Pitfield, Milton Keynes, MK11 3LW, UK
UKHW022158190926
475702UK00003B/86
9781138059351